T0234755

T-Labs Series in Telecommunication Services

Series Editors

Sebastian Möller, Quality and Usability Lab, Technische Universität Berlin, Berlin, Germany

Axel Küpper, Telekom Innovation Laboratories, Technische Universität Berlin, Berlin, Germany

Alexander Raake, Audiovisual Technology Group, Technische Universität Ilmenau, Ilmenau, Germany

More information about this series at http://www.springer.com/series/10013

Patrick Ehrenbrink

The Role of Psychological Reactance in Human–Computer Interaction

 Springer

Patrick Ehrenbrink
Werder (Havel), Brandenburg, Germany

Zugl.: Berlin, Technische Universität, Diss., 2018

ISSN 2192-2810   ISSN 2192-2829   (electronic)
T-Labs Series in Telecommunication Services
ISBN 978-3-030-30312-9   ISBN 978-3-030-30310-5   (eBook)
https://doi.org/10.1007/978-3-030-30310-5

This Springer imprint is published by the registered company Springer Nature Switzerland AG
The registered company address is: Gewerbestrasse 11, 6330 Cham, Switzerland

Preface

This book is titled "The Role of Psychological Reactance in Human–Computer Interaction". In order to appropriately investigate this topic, three main reseach questions were formulated.

1. Is Psychological Reactance Relevant for Human–Computer interaction?
2. How can State Reactance be Measured?
3. What Factors Influence State Reactance?

The book is divided into five parts. Part I contains the motivation for the book, as well as the state of the art chapter.

Classical usability tests often rely on self-report measurements about the user's judgment of an examined system and behavioral observations, often aimed at measuring performance and identifying usability problems. However, there are certain effects that are especially hard to detect in the frame of a classical user test, namely psychological effects that might address and trigger mechanisms which are not usually known by usability and user experience experts. One of such effects is psychological reactance. This book examines the role of psychological reactance within the field of human–computer interaction.

The book starts with a motivation chapter, explaining why psychological reactance might be of importance for human–computer interaction. It claims that the recent rise of intelligent and connected devices leads to situations in which users cannot fully understand the functionality of some technical devices or even get overwhelmed while using them. The associated reduction of control over the devices might trigger state reactance because state reactance is a motivational state which results from a perceived loss of freedom of control.

After the motivation, the state of the art in psychological reactance research is presented. Psychological reactance can be divided into two concepts. Trait reactance, which is a personality trait, and state reactance, which is a motivational state. The concept of state reactance is explained in further detail, including a set of potential moderators that might determine the strength of the reactance consequences. Also, means of measuring state and trait reactance are presented and discussed. It is concluded, that there is no adequate measurement technique for state reactance, which could be used in the current work.

The main goal of Part II is to find out if psychological reactance is of relevance to the field of human–computer interaction. A two-fold strategy was applied to gather data about psychological reactance in human–computer interaction. One strategy was to identify relevant scientific literature via a literature search. The other strategy was running a survey among usability experts and ask them about their experiences with state reactance. The results of both strategies were analyzed in terms of what situations in human–computer interaction triggered state reactance, and what were the consequences. The literature search identified seven groups of situations where reactance in connection with Human–Computer Interfaces was investigated. The expert survey produced five groups of situations in which state reactance was triggered, all of these groups but "system errors" were covered by literature. Therefore it was decided to conduct an experiment that was dedicated to system errors and their implications for state reactance.

Also, it was found that the evidence for trait reactance being of relevance for human–computer interaction is quite scarce. A Study involving a Smart Home System was conducted, which indicated that trait reactance can have an influence on how adaptive behavior of smart homes is perceived.

Before state reactance could be investigated further, an adequate measurement technique had to be developed. The construction of the Reactance Scale for Human–Computer Interaction (RSHCI) is described in Part III. The development of the RSHCI involved collecting phrases from anonymous user comments on websites because these were deemed to be especially valid. Also, an online study was conducted to collect a large set of data for the factorial analysis. The factorial analysis involved a maximum-likelihood factor analysis approach, followed by a confirmatory analysis, using structural equation modeling. Its criterion validity was shown in a subsequent validation study.

Part IV investigates the research question regarding the factors that influence psychological reactance. The part involves a closer examination of questions that are raised fbyrom Part II. At first, the reports by the usability experts, claiming that system errors can cause state reactance was investigated with a laboratory experiment involving a smart TV system. The results show that system errors can cause state reactance, but that appropriate explanation of the cause of the error can reduce this effect significantly. The next study investigated if and how the potential moderator variables that have been described in the state of the art section, influence the components of state reactance and its effect for the users' global judgment of the system. It was also shown that state reactance can increase the prediction accuracy of the global judgment of the system.

Part V then discusses the general implications of the results presented in this book and concludes with an outline of open questions and future work.

Werder (Havel), Germany   Patrick Ehrenbrink

Acknowledgements

This work was supported by the Bundesministerium für Wirtschaft und Energie (Germany) under Grant no. 01MG13001G Universal Home Control Interfaces @ Connected Usability (UHCI) and by the Bundesministerium für Bildung und Forschung (Germany), Software Campus program, Grant no. 1043569, Sozialpsychologische Aspekte von Smart Homes (SPASH).

The work that has been conducted to write this book would not have been possible without the many people who have supported me over the last years. First and foremost, I want to thank my supervisor and friend, Sebastian Möller, who has always provided academic guidance and showed endless patience toward me finding my topic. Also, thank you to my co-supervisors Khin Than Win and Sergio Lucia.

The administrative team of the chair deserves also special thanks. Irene Hube-Achter was the great enabler at the chair and solved countless problems that I encountered during the years. Yasmin Hillebrenner was always there to help when I needed last minute vacation or advice and Tobias Hirsch could provide most gadgets that one would ever need for a Dissertation about HCI.

I also would like to thank my colleagues, former colleagues, and friends at work: Stefan Hillmann, Benjamin Weiss, Maija Poikela, Rahul Swaminathan, Babak Naderi, Tim Polzehl, Jan-Niklas Voigt-Antons, Britta Hesse, Neslihan Iskender, Tanja Kojic, Thilo Michael, Gabriel Mittag, Falk Schiffner, Carola Trahms, Stefan Uhrig, Saman Zadtootaghaj, Rafael Zequeira, Benjamin Bähr, Justus Beyer, Klaus-Peter Engelbrecht, Ina Wechsung, Lydia Kraus, Friedemann Köster, Marie-Neige Garcia, Dennis Guse, Marc Halbrügge, Florian Hinterleitner, Tilo Westermann and many more. I have always felt comfortable at the QU labs and without the lunch and coffee breaks and kicker sessions, I sure would have missed something.

During the years I also had a lot of support from people who contributed to my research either as student workers or study projects and theses. I would therefore like to thank Lea Soldo, Duy Phuong Bui, Fabienne Roche, George Göcksel, Xin Guang Gong, Seif Osman, Elvira Ibragimova, and Christian Krüger.

It was a long journey through many institutions for me to finally complete this book and I could always trust that my parents Marlies Ehrenbrink and Ludger Ehrenbrink, as well as my sister Elena Ehrenbrink would be there to support me. Thank you!

My daughter Maja was born while I was still working on my dissertation. She was and will be the sunshine of my life and I am grateful for that. In the end, I would like to thank Sabine Prezenski. You and I have been through ups and downs of life and work. But no matter how tricky the situation was, you were always there to lead the way. I would not be who I am today without you and your coaching. I am happily looking forward to spending my life with you!

Contents

Acronyms

AAL   Ambient Assisted Living
AGFI   Adjusted Goodness of Fit Index
CFI   Comparative Fit Index
df   Degrees of freedom
FBMPS   Fragebogen zur Messung der Psychologischen Reaktanz
HPRS   Hong's Psychological Reactance Scale
P.851   A questionnaire according to ITU-T recommendation P.851 [1, 2]
RMSEA   Root Mean Square Error of Approximation
RSS   Reactance Restauration Scale
RSHCI   Reactance Scale for Human-Computer Interaction
SAM   Self Assessment Manikin
SSRS   Salzburger State Reactance Scale
TRS   Therapeutic Reactance Scale
TV   Television set

References

1. International Telecommunication Union: Subjective quality evaluation of telephone services based on spoken dialogue systems. Tech. Rep. Supplement 851 to P-Series Recommendations, International Telecommunication Union, Geneva, Switzerland (2003)
2. Möller, S., Smeele, P., Boland, H., Krebber, J.: Evaluating spoken dialogue systems according to de-facto standards: A case study. Comput. Speech Lang. **21**, 26–53 (2007)

Part I
Theory

Chapter 1
Introduction

Classical usability and user-experience testing involve a thorough and systematic evaluation of the system that is under evaluation. This may be done with user-centric, or expert-centric methods [10, p. 24]. The goal of such an evaluation is usually to test if a system acts as expected and meets the requirements that its users have [8, p. 319].

In recent years, services and connected devices such as smart homes and intelligent personal assistants, e.g., Siri [4], have become available to the consumer market. The first version of Siri was introduced in 2011, one year later, Google released the first version of Google Now [3] and Amazon released Alexa in 2014 [3]. Those intelligent personal assistants are cloud services that can be updated by the providing companies without the knowledge or approval of their users. Since their introduction to the market, there have been many updates to the assistants that added new features, for example, the possibility to control smart home devices [1, 5, 17]. Other examples of such newly added features include support for music-streaming services, answering almost arbitrary questions about facts or ordering pizza.

For the users of such systems, the ever-increasing set of functionalities might be useful and welcome in many situations. However, it also means that the systems get very complex. Besides the complexity of such systems, the fact that features are added in the cloud also means, that it is almost impossible for a user of such a system to fully learn all its functionalities and to understand the mechanics behind those functionalities. And, even if a user would manage to learn all functions that are available at some point via a documentation, he or she would not be certain if the system would have the same range of functionality the day after. As a result, users will not become experts for such systems and might even be overwhelmed while using them.

Of course, developers are aware of such problems and have come up with solutions to improve user experience and facilitate interactions with such systems. For example, intelligent personal assistants are capable of processing commands in natural

© Springer Nature Switzerland AG 2020
P. Ehrenbrink, *The Role of Psychological Reactance in Human–Computer Interaction*, T-Labs Series in Telecommunication Services, https://doi.org/10.1007/978-3-030-30310-5_1

language to some extent (e.g., Siri, Cortana, Alexa, Google Now), thereby dramatically reducing path lengths in menus. Also, many services are able to act proactively, for example by offering alternatives if a search query did not produce results, such as google.com. Another famous example of a proactive system was the Office Assistant in Microsoft Office. It contained a personal assistant called Clippit, who would often show up in Microsoft Word and offered help for writing a letter or offer some sort of assistance after anticipating what the user's intention might have been.[1] Clippit was not well received by the users of Microsoft Office. A fact that Microsoft referred to, by calling it *infamours*, when it announced that Clippit would be removed from future versions of Microsoft Office [15].

If users do not rationally understand how a system works, what it can do and where its limitations are, they might cope with that situation by relying on interaction schemes that they know from interaction with humans. Already Nass and Moon found that, in a human–computer interaction scenario, humans can fall into interaction schemes that are known from human–human interaction, such as politeness [16]. Cowan and Branigan found that humans can fall back into using interaction schemes from human–human interaction when interacting with computer systems, by using lexical alignment when interacting with spoken dialogue systems [7].

Proactive interaction of any technical systems with humans usually requires that the system anticipates the user's actions or intention. Based on those assumptions, the system can then propose an action, information or a product of which it thinks that the user might be interested in. Such proposals are shortcuts to a specific action, information or product, making them easier available. However, this also implies that alternatives will become relatively hard to reach (compared to the suggested one), which could be regarded as a behavioral restriction or paternalism by the user. As will be explained in Sect. 2.2, behavioral restrictions, or more generally, a perceived loss of freedom, can lead to a state of reactance. Reactance is a concept from social psychology. If someone feels her or his personal freedom to be under threat, that person might become reactant. Reactance can result in decreased acceptance of the source of the perceived threat to freedom. It would, therefore, be undesirable for the creators of technical systems, if users would become reactant while using these systems, provided that humans can become reactant from interacting with technical systems.

As it was explained before, users can fall back into interaction schemes from human–human interaction when interacting with technical systems. Falling back into such schemes is probably facilitated when the behavior and appearance of the technical systems become more human-like. It was already mentioned, that intelligent systems may rely on natural language interaction and proactive behavior, what can resemble human behavior. But there are also other aspects in which intelligent systems are becoming more humanlike. The intelligent personal assistant Cortana [14] can tell jokes and Siri [4] can use sarcasm and humor to answer inappropriate or senseless questions (see Section for example dialogues.). Taken together, all these

---

[1] In fact, interaction with Clippit might be a common situation for many users, where they experienced state reactance.

features that used to be exclusive to the domain of human–human interaction and are now becoming present in the domain of human-computer interaction, appear to be useful in keeping users engaged with systems, even if they don't fully know if they will be helpful for a task, or not. But at the same time, higher human likeness might facilitate falling back into interaction schemes from human–human interaction. Due to the likely loss of control and autonomy of the users in some situations, a strong candidate for such a scheme is psychological reactance.

Scope

After arguing why psychological reactance might be of interest for intelligent systems, the next sections will get into more detail about what kinds of technical systems might be especially prone to trigger psychological reactance, and why.

Intelligent Systems

It was argued before, that intelligent systems like intelligent personal assistants might be more capable of triggering psychological reactance than non-intelligent systems like a toaster. This is based on the assumption that their functionality suggests intelligent or agenda-driven behavior, which marks them as a socially relevant agent. In order to further clarify this argumentation, it is important to define the notion of an intelligent system. Literature knows some definitions, such as the one provided by Möller et al., who described intelligent devices via the ability to maintain a spoken dialogue interaction on their own [13]. Such definitions are usually formulated with an emphasis on the current domain. Therefore, a custom definition of intelligent systems will be introduced here:

> An intelligent system is a technical system that functions in a way that it suggests agenda driven behavior. At the same time, the execution of such behavior is salient to the user of the intelligent system.

The impression of agenda-driven behavior is an important part of the definition because of psychological reactance being a social phenomenon. It was suggested, that psychological reactance cannot be triggered in a "noninterpersonal" situation [9]. This is in line with the experiments by Roubroeks et al. [19–21], who found that participants showed a higher state of reactance when social agency was high, compared to when it was low. Agenda-driven behavior is one characteristic of social agents, such as humans. The perception of such a characteristic might be sufficient to trigger the formerly mentioned schemes from human–human interaction and among them, psychological reactance. In general, such schemes could also be anything other than reactance, such as the above-mentioned politeness [16] or lexical allignment [7]. However, there are some classes of intelligent systems that pose a threat to the freedom of choice or control to the users and the most likely scheme that could be triggered by such threats is psychological reactance. Therefore, some classes of intelligent systems that are especially vulnerable for triggering psychological reactance will be mentioned in the following paragraphs.

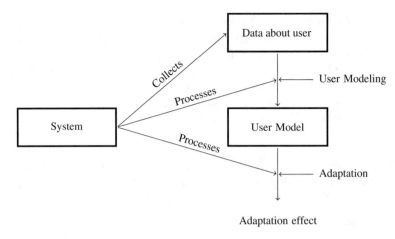

Fig. 1.1 Model of an adaptive system, from [6]

Adaptive Systems

In recent years, adaptivity of technical systems has been a topic of interest among scholars [12, 18]. This might be partly due to the employment of artificial intelligence and cognitive approaches, which are necessary to accomplish meaningful adaptation. Adaptation is usually performed by some form of system intelligence which applies a set of rules on the basis of collected data from the user or the environment. Specht describes adaptation as being a process consisting of a diagnose and an adaptation process [22]. Brusilovsky published a more detailed definition about user-adaptive systems [6]. His model of a user-adaptive system is shown in Fig. 1.1. It involves user data that is used to build a user model. This user model is then used to decide on the adaptation process. Instead of user-data, one could also build a system that uses data about the environment (or data about the environment and the user and some other context) in a world model to be able to choose more informed adaptation strategies.

For some systems, adaptation is rather subtle, e.g., if a laptop adjusts its CPU frequency to the task that it is performing. Adaptive systems that only employ such subtle adaptation are not in the scope of this book because their adaptative behavior is not salient enough to pose a threat to the user's freedom of control. Adaptive systems are often adaptive to increase the ease of use for their users. A good example of this is a smartphone. Smartphones need to be usable indoors, but also in bright sunlight. As interaction is mostly accomplished via touchscreens, adjusting the screen brightness requires users to be able to see what is on the screen. In bright sunlight, the content of the screen is usually not visible. Therefore, adaptive screen brightness constitutes a big improvement in usability[2] and ease of use[3] because users would not be able

---

[2]"The extent to which a system, product or service can be used by specified users to achieve specified goals with effectiveness, efficiency, and satisfaction in a specified context of use" [11].

[3]Ease of use is a measurement of how easy a technical system is to use by its users.

to turn the screen brighter if they could not see the controls for it on the screen. Adaptation might often save time and effort for users, but it also necessarily takes away some control of the device's functionality from the users, which might trigger psychological reactance.

Ubiquitous Systems and Cloud-Based Systems

Ubiquitous systems are interconnected systems that are also known as smart things. The concept of ubiquitous computing was introduced by Weiser [23]. One form of a ubiquitous system is a smart home system. Such systems are often a collection of connected devices that are able to communicate with one another. This communication is not necessarily done via local networks, but usually via the internet. Also, such devices are often merely client devices, the actual service is a cloud service. For example, this is the case for smart speakers like the Echo by Amazon.com [2]. Ubiquitous systems can have the advantage, that they can be extended with more devices. Such groups of devices can then build a smart home or a smart office. They can be controlled via a variety of modalities and interfaces, such as speech commands directed at an intelligent personal assistant [1, 17], via smartphone apps or via remote controls. Also, since their software is often implemented as a cloud service, they can always receive the newest updates and functions, often without the explicit consent or even knowledge of the user. This alleged ease is again a potential threat to the user's freedom of being in control of the devices and therefore a likely trigger for reactance.

Motivation

New types of devices and services, such as intelligent and ubiquitous system are in danger of eliciting a perception of being a threat to the freedom of being in control to their users. A circumstance that can result in the users becoming reactant. As will be explained in Sect. 2.2.1.1, psychological reactance can have serious consequences for the system that triggers it, including lower acceptance and even hostile behavior of the user. The aim of this book is to investigate the role of psychological reactance for human–computer interaction. Therefore, it shall be investigated in what situations psychological reactance is of relevance for human–computer interaction, how it can be measured and what can be done to prevent or alleviate its consequences.

References

1. Amazon.com, Inc.: Amazon echo: Always ready, connected and fast (2016). Retrieved February 05, 2016 from http://www.amazon.com/Amazon-SK705DI-Echo/dp/B00X4WHP5E# compare
2. Amazon.com, Inc.: Amazon echo—black (1st generation) (2018). Retrieved August 19, 2018 from https://www.amazon.com/Amazon-Echo-Bluetooth-Speaker-with-Alexa-Black/dp/B00X4WHP5E

3. Amazon.com: Alexa (2014). Retrieved January 27th, 20117 from https://developer.amazon.com/alexa

4. Apple Inc.: Siri (2011). Retrieved December 23, 2016 from http://www.apple.com/ios/siri/

5. Apple Inc.: Use homekit-enabled accessories with your iphone, ipad, and ipod touch (2016). Retrieved February 05, 2016 from https://support.apple.com/en-us/HT204893

6. Brusilovsky, P.: Methods and techniques of adaptive hypermedia. User Model. User-Adapt. Interact. **6**, 87–129 (1996)

7. Cowan, B.R., Branigan, H.P.: Does voice anthropomorphism affect lexical alignment in speech-based human-computer dialogue? In: Sixteenth Annual Conference of the International Speech Communication Association (2015)

8. Dix, A., Finlay, J., Abowd, G.D., Beale, R.: Human-Computer Interaction. Prentice Hall, Pearson (2004)

9. Heilman, M.E., Toffler, B.L.: Reacting to reactance: an interpersonal interpretation of the need for freedom. J. Exp. Soc. Psychol. **12**(6), 519–529 (1976). https://doi.org/10.1016/0022-1031(76)90031-7. URL http://www.sciencedirect.com/science/article/pii/0022103176900317

10. Informationstechnische Gesellschaft im VDE (ITG): Messung und Bewertung der Usability in Smart Home-Umgebungen. Tech. Rep. ITG 2.1-02, Informationstechnische Gesellschaft im VDE (ITG) (2014)

11. International Organization For Standardization: ISO 9241–11—Ergonomic Requirements for Office Work with Visual Display Terminals (VDTs): Part 11: Guidance on Usability. ISO, Geneva (1998)

12. Jameson, A., Gajos, K.Z.: Systems that adapt to their users. Fundamentals, Evolving Technologies and Emerging Applications. CRC Press, Boca Raton, FL, The Human-Computer Interaction Handbook (2012)

13. MS.: Evaluating the speech output component of a smart-home system. Speech Commun. **48**(1), 1–27 (2006). https://doi.org/10.1016/j.specom.2005.05.004. URL http://www.sciencedirect.com/science/article/pii/S016763930500124X

14. Microsoft: Cortana (2014). Retrieved December 23, 2016 from https://www.microsoft.com/windows/cortana/

15. Microsoft: Farewell clippy: whats happening to the infamous office assistant in office xp (2001). Retrieved August 6th, 2018 from https://news.microsoft.com/2001/04/11/farewell-clippy-whats-happening-to-the-infamous-office-assistant-in-office-xp/

16. Nass, C., Moon, Y.: Machines and mindlessness: social responses to computers. J. Soc. Issues **56**(1), 81–103 (2000). https://doi.org/10.1111/0022-4537.00153. URL https://spssi.onlinelibrary.wiley.com/doi/abs/10.1111/0022-4537.00153

17. Nest Labs: How do google voice actions and google now work with the nest thermostat? (2016). Retrieved February 09, 2016 from https://nest.com/support/article/How-do-Google-voice-actions-and-Google-Now-work-with-the-Nest-Thermostat

18. Park, J., Han, S.H.: Integration of adaptable and adaptive approaches for interface personalization through collaborative menu. Int. J. HumanComputer Interact. **28**(9), 613–626 (2012). https://doi.org/10.1080/10447318.2011.653325

19. Roubroeks, M., Ham, J., Midden, C.: The dominant robot: Threatening robots cause psychological reactance, especially when they have incongruent goals. In: Ploug, T., Hasle, P., Oinas-Kukkonen, H. (Eds.) Persuasive Technology, pp. 174–184. Springer, Berlin Heidelberg (2010)

20. Roubroeks, M., Ham, J., Midden, C.: When artificial social agents try to persuade people: the role of social agency on the occurrence of psychological reactance. Int. J. Soc. Robot. **3**(2), 155–165 (2011). https://doi.org/10.1007/s12369-010-0088-1

21. Roubroeks, M., Midden, C., Ham, J.: Does it make a difference who tells you what to do ? exploring the effect of social agency on psychological reactance. In: Proceedings of the 4th International Conference on Persuasive Technology, Persuasive '09, pp. 15:1–15:6. ACM, New York, NY, USA (2009). https://doi.org/10.1145/1541948.1541970. URL http://doi.acm.org/10.1145/1541948.1541970

22. Specht, M.: Adaptive methoden in computerbasierten lehr/lernsystemen. Ph.D. thesis, Universität Trier (1998)
23. Weiser, M.: The computer for the 21st century: specialized elements of hardware and software, connected by wires, radio waves and infrared, will be so ubiquitous that no one will notice their presence. In: Baecker, R.M., Grudin, J., Buxton, W.A.S., Greenberg S. (Eds.) Readings in Human-Computer Interaction, Interactive Technologies, pp. 933–940. Morgan Kaufmann (1995). https://doi.org/10.1016/B978-0-08-051574-8.50097-2. URL http://www.sciencedirect.com/science/article/pii/B9780080515748500972

Chapter 2
State of the Art

2.1 Introduction to Psychological Reactance

From the beginning of reactance research, psychological reactance was conceptualized as a state. In his first book about the topic [3, p. 2], Brehm described psychological reactance:

> Since this hypothetical motivational state is in response to the reduction (or threatened reduction) of one's potential for acting, and conceptually may be considered a counterforce, it will be called "psychological reactance".

The concept of psychological reactance as a personality trait only emerged several years later. Accordingly, state reactance will be introduced in detail in the following sections. Trait reactance will then be introduced afterward.

2.2 State Reactance

When Brehm first introduced the concept of psychological reactance, he described it as a motivational state. He wrote that persons, who feel their behavioral freedoms being under threat, will become motivationally aroused, and that this arousal will eventually be directed toward reestablishment or protection of the freedom that is under threat [3, p. 2]. This mechanism is called psychological reactance and the motivational state that a person can experience when confronted with a threat to their freedom is called state reactance. Since its introduction in 1966, psychological reactance has been investigated mainly in the context of social psychology and marketing.[1] There have also been some attempts to apply reactance theory to findings of human–computer interaction research (see Chap. 4), but a systematic evaluation of its validity for the use in this domain has not been attempted.

---

[1]For an overview of reactance research refer to [36].

© Springer Nature Switzerland AG 2020
P. Ehrenbrink, *The Role of Psychological Reactance in Human–Computer Interaction*, T-Labs Series in Telecommunication Services,
https://doi.org/10.1007/978-3-030-30310-5_2

As mentioned before, state reactance is usually triggered when a person perceives a threat (or a real reduction) of exercising his or her freedom of choice or freedom of being in control [3, p. 2]. In a book published in 1981 [4], Brehm and Brehm specified the notion of freedom further, arguing that the threatened freedoms need to be specific [4, p. 12]:

> ...the freedoms addressed by the theory are not "abstract considerations," but concrete behavioral realities. If a person knows that he or she can do X (or think X, or believe X, or feel X), then X is a specific, behavioral freedom of that person.

For the current work, this means that also the situations, in which technical systems might trigger state reactance, need to be situations in which the user is restricted or threatened in a freedom that he or she believed to be desirable or realistic to really execute it. This means that e.g., a collision avoidance system in a car might restrict the driver's freedom to crash into another car, but since this behavior is not desirable to be executed, the collision avoidance system would rather not trigger state reactance. On the other hand, an assistance system that prevents drunken or tired drivers from starting the car would likely cause state reactance, because, in such a situation, the driver really wants to execute the freedom of driving the car.

When a person experiences a restriction of a threat to freedom, state reactance will lead to the motivation of restoring the lost freedom. In order to release reactance arousal, e.g., by accomplishing reinstatement of the lost freedom, multiple strategies are possible. These strategies are often undesirable for the source of the freedom threat, or the creators of technical systems that trigger state reactance. But at the same time, they can be used as indicators for state reactance.

2.2.1 Consequences of State Reactance

Preservation of Other Freedoms

One strategy of releasing reactance arousal is by protecting the behavioral freedoms that are not jet restricted. Brehm and Brehm wrote, that a person who employs this strategy might experience even more reactance the next time when a now more highly valued freedom is under threat [4, p. 115].

Denial of Threat

Another strategy of releasing reactance arousal is a denial of the freedom threat. This strategy is usually not a viable option because ignoring the threat is either not well received in social situations [4, pp. 111], or not possible in the context of human–computer interaction, because the controls for still executing the denied functionality are just not there, anymore. Ignoring the threat is, however, often possible when it is posed by an assistant system, e.g., via a suggestion or via proactive action that can be overwritten by the user.

2.2.1.1   Indicators of State Reactance

Before the introduction of appropriate metrics for state reactance, scholars relied on behavioral observations to draw a conclusion about if a person experienced a state of reactance, or not. If a participant would behave in a way that was expected of a person experiencing reactance, that is by applying certain strategies of releasing reactance arousal, one could conclude that this participant would really experience reactance. Two prominent behavioral patterns that have been used as indicators for state reactance are source derogations and boomerang effects.

Source Derogation

Source derogation is a behavior often observed in persons experiencing state reactance [17, 29]. If a person experiences a freedom threat, one strategy of trying to reestablish the lost freedom can be downgrading the source of the threat. For example, Grandpre et al. conducted an experiment where children received persuasive messages in videos and found that the video quality of explicit messages (high threat to freedom) was rated worse than the video quality of implicit messages (lower threat to freedom) [17]. The more negative rating of the video quality can be interpreted as an attempt to reestablish freedom by derogating the source of the threat.

Boomerang Effect

Another indicator of state reactance which is often used is the boomerang effect. A boomerang effect means that a person changes its behavior in a way that it contradicts the freedom threat, hence the person tries to reestablish the lost freedom by acting against the threat or its perceived intention [4, pp. 98]. An example of this is a child that can choose between a glass of juice and a glass of milk. It might be in favor of having the glass of milk, but then its father tells the child that it should really not have the juice because it is unhealthy. The child might perceive its father's hint as a threat to its freedom of choice and then become reactant. This might result in a boomerang effect and instead of choosing the glass of milk (as intended before), it chooses the glass of juice, just to reinsure its freedom of choice. Boomerang effects are often encountered in public health campaigns. E.g. Burgoon et al. found that students who are exposed to anti-drug campaigns with strong controlling language (see Sect. 2.2.3) show more intend to use drugs in the future than others [6, p. 80].

2.2.2   The Intertwined Process Cognitive-Affective Model

The concept of psychological reactance has been formed already in 1966 [3], but there has been a lack of a precise conceptualization of its cognitive mechanisms [37]. In 2005, Dillard and Shen [10] were the first researchers trying to establish a precise model of the cognitive mechanisms of state reactance [37]. In their original paper, they compared four different models of state reactance [10] that use anger, negative cognitions or both for the conceptualization of state reactance. The concept of anger and negative cognitions as parts of reactance has already been introduced by Brehm. He wrote that a great magnitude of state reactance might lead to hostile and aggressive feelings [3, p. 9] (anger). State reactance might also lead to an attempt to restore the lost freedom by downgrading the source of the freedom threat (source derogation) [45], which might occur in the form of negative thoughts or cognitions directed at the source of the freedom threat. To illustrate the possible mechanisms behind state reactance, the four models of Dillard and Shen [10] and one model by Rains [37], will be described shortly. The first two models of the list are single process models that assume that state reactance can be seen as either a solely cognitive, or a solely affective process. The next three models are more complex and depict state reactance as a construct consisting of cognitive and affective components.

Single Process Cognitive Model

The single process cognitive model follows the assumption that state reactance is only a cognitive phenomenon, rather than an affective one. An argument for this model is that state reactance results in negative cognitions, including counterarguing to a perceived threat or downgrading it. This behavior can be interpreted as an attempt to reestablish the lost freedom by either arguing against it (counterarguing) or decreasing its importance (downgrading). Therefore, the model assumes that a freedom threat will cause negative cognitions, which will then result in a change of attitude and behavior.

Single Process Affective Model

The single process affective model follows the assumption that state reactance is only an affective phenomenon, consisting of anger. This follows the description by Brehm [3] that Reactance results in aggressive feelings. Dillard and Shen argue, that one could see state reactance as belonging to the family of Anger, similar to irritation, annoyance, and range [10]. The single process affective model assumes that a freedom threat can cause Anger, which will then be followed by a change of attitude and behavior.

Dual Process Cognitive-Affective Model

The dual-process model assumes that there are two distinct processes that affect behavior. These two processes are a cognitive process and an affective process, both influencing attitude and behavior independently.

Intertwined Process Cognitive-Affective Model

The final model that Dillard and Shen proposed was the intertwined process cognitive-affective model. It follows similar assumptions that the dual-process cognitive-affective model. The difference is that this model does not regard the affective and the cognitive mechanisms as independent from each other, but assumes that both are intertwined considerably. The intertwinement is to such a degree, that the individual influence of either of the two processes on attitude and behavior cannot be predicted, only their combined influence can.

Dillard and Shen used structural equation modeling, a mathematical method which allows for the testing of computational models on empirical data [7], to compare the four models based on empirical data from two studies [10]. Their results led them to the conclusion that the intertwined process cognitive-affective model would fit the observed data best.

Linear Affective-Cognitive Model

In a subsequent analysis, Rains compared the dual- process cognitive-affective model and the Intertwined Process Cognitive-affective Model against the linear affective-cognitive model [37]. The linear affective-cognitive model is similar to the two other models in that it also regards state reactance as a construct consisting of anger and negative cognitions, but it also assumes, that the affective component anger precedes the cognitive component negative cognitions. Like Dillard and Shen [10], also Rains [37] compared the models using empirical data from two different studies. Rains found that the intertwined process cognitive-affective model fits the observed data best. Rains later also conducted a meta-analytic review of reactance research to compare the three models linear affective-cognitive model, dual process cognitive-affective model and intertwined cognitive-affective model on a large set of accumulated data (N = 4942) [36]. Again, Rains concluded that the intertwined process cognitive-affective model fits the observed data best [36]. The model is depicted in Fig. 2.1.

Criticism of the Intertwined Model

Steindl et al. point out that the intertwined model might not generalize to all situations where state reactance occurs because there might be two processes involved: One, where persons enter a reactant state directly after the perceived freedom threat and one where persons show a delay before entering the reactant state [44, 46]. This delay might be due to intermediary cognitive processes [44, 46]. The intertwined model does not account such intermediary processes and might therefore not be valid in situations where state reactance occurs only after some delay. One situation in which this occurs is when state reactance is aroused by the experience of legitimate freedom threats [44].

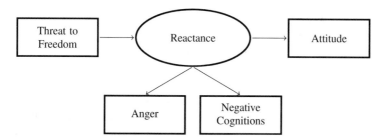

Fig. 2.1 The intertwined model of state reactance. A threat to freedom will cause state reactance, which will then change the Attitude of a Person. Anger and negative cognitions toward the freedom threat indicate Reactance

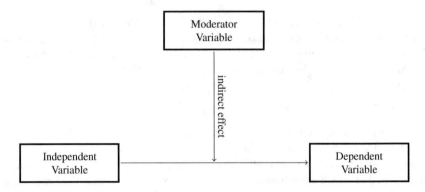

Fig. 2.2 Influence of a moderator variable

2.2.3 Moderators of State Reactance

There is a number of possible moderators for state reactance. A moderator variable is a function or characteristic of a technical system, the user, or the environment, that influences the effect of an independent variable on the dependent variable [1, pp. 682]. The effect of a moderator variable is called interaction effect [15, 395]. The interaction effect is an indirect effect of a moderator variable on a dependent variable. This mechanism is illustrated in Fig. 2.2.

Literature knows several moderators of state reactance. In this work, three of these will be discussed because of their potential relevance for human–computer interaction in general, and for adaptive services especially. The moderators are trait reactance, social agency, and involvement. They will be described in the following paragraphs.

Trait Reactance

The most obvious moderator of state reactance is trait reactance. Trait reactance is a personality trait that determines how prone a person is of becoming reactant.[2] In theory, a person's level of trait reactance could, therefore, be used to predict how likely that person is of becoming state reactant or how strong that person's reactance-reaction might be. However, research has led to mixed results in this regard. Quick and Stephenson found that persons with a high value in trait reactance show a higher correlation of perceived threat and state reactance than persons with a low value in trait reactance, but a closer investigation of the two components of state reactance, anger, and negative cognitions revealed that only anger showed a correlation with trait reactance, not negative cognitions [32]. Also, Dillard and Shen observed a correlation between trait reactance and state reactance in an experiment where they used texts about flossing teeth as stimuli. The correlation could not be found in a similar experiment, where texts about alcohol consumption were used. They concluded, that the topic itself might be responsible for the different results [10].

---

[2]For a deeper introduction see Sect. 2.3.

Trait reactance is of relevance for human–computer interaction because it is a personality trait. Personality traits are characteristics of persons that are relatively stable over time [31, p. 541] and can hence be used as predictors for personality-adaptive services [14]. A personality trait has to be assessed only once and can afterward be relied on for many years.

Social Agency

Social agency theory claims that social cues in multimedia messages can prime a social conversation schema in people [28]. This means that e.g., a participant that is interacting with a computer via a chatbot might start to act (to a limited extent) as if the computer was a real, social interaction partner if the computer uses social cues while communicating. Therefore, if a computer uses social cues while interacting with humans, it might become a social agent. The social agency theory is supported by Nass and Moon, who conducted a literature review and showed that people tend to treat computers as social agents by applying social norms and expectations to them [30].

Social agency is a moderator variable that is relevant for the current work because of psychological reactance being a concept from social psychology. As such, one could assume that state reactance is stronger if the source of the freedom threat is a social agent [39] or if the situation is of social relevance. Therefore, scholars started to investigate if a freedom threat triggers more state reactance if it originates from a source that appears like a human or at least a social agent. One of the first experiments to investigate this was done by Roubroeks et al. They conducted an online experiment where they exposed participants to high threat and low threat messages and accompanied those messages with either a still image or an animated video showing a robotic cat. They found that participants who were exposed to the animated video experienced higher levels of state reactance than the participants who were exposed to the still images (and the baseline condition without any images). This was true for the low threat condition and the high threat condition [39, 40]. A replication of the experiment in the laboratory confirmed that high social agency is associated with higher state reactance [38]. Also, Ghazali et al. conducted an experiment in which they varied social Agency of a robot with a variety of other variables (controlling language and involvement). They observed higher state reactance when the robot had a higher level of social agency, as well [16].

Social agency is relevant for adaptive services and human–computer interaction because it is thought to be a system feature which influences users' perception of systems [8]. An interactive system can, for example, hint at its capabilities by adjusting social agency of its avatar. If social agency really causes state reactance to be stronger, as was found in [39, 40], then using agents, capable of social cues could easily backfire and methods of easing state reactance levels would have to be found and applied.

Involvement

In their experiment, Ghazali et al. also found that task involvement has an effect on state reactance. They observed higher state reactance if the participants were more involved in their tasks [16]. Similar effects were predicted by Brehm, who stated that the amount of Reactance that a person will experience is dependent on how important the issue in question is to the subject [3, pp. 118]. On the other hand, Quick investigated the effect of issue involvement on state reactance and did not observe any direct influence of involvement on state reactance [35].

Involvement as a moderator variable for state reactance can become relevant for human–computer interaction situations in which a system has to decide on what tasks to prioritize, e.g., an automatic task management system of a personal computer could avoid state reactance by prioritizing processes that are more important to the user and thereby help to maintain higher acceptance levels for the device. Of course, such a system would presume near-perfect accuracy, since a wrong prioritization could result in even more state reactance.

2.2.3.1   Other Moderators

Literature has also named other moderators of state reactance, namely magnitude of request and severity of consequences. A high magnitude of request means that a person that receives a request has to invest a large amount of effort, time or other resources to fulfill it [37]. Hence, the person's freedom is threatened by the amount of resources that have to be provided to fulfill the task. If the request is large enough, the involved person might enter a reactant state. Magnitude of request has been used as an independent variable for stimuli in state reactance research to generate a freedom threat [37]. With increasing resources that are required to fulfill a request, also the ability to allocate these resources to exercising personal freedoms decreases and the freedom threat emerges. Severity of consequences follows the same argument. In fact, a higher magnitude of required resources can be regarded as a more severe consequence. Severity of consequences and magnitude of request will not be further investigated in this book because there is already a large body of evidence that more severe consequences or a higher magnitude of request can increase state reactance levels or decrease user acceptance [10, 16, 37].

2.2.4   State Reactance Assessment

There have been several attempts to assess state reactance, so far. A literature search revealed four measurement tools that claim to measure state reactance. These methods will be introduced briefly in the following paragraphs.

Dillard and Shen Method

Out of the four measurement tools, a mixed-method approach introduced by Dillard and Shen [10] is the one most widely used. It follows the intertwined model of state reactance and uses a four-item questionnaire for anger on a 5-point scale ranging from 0 = "none of this feeling" to 4 = "a great deal of this feeling" [10]. The items are: "irritated", "angry", "annoyed", and "aggravated". Negative cognitions are assessed using thought listing. In this technique, participants write down all thoughts that come to their mind. Afterward, the thoughts are reviewed and segmented into thought units. All thought units that represent affect are removed in the next step, as those would be redundant with the measurement of anger. Then, all units that are not relevant for the topic are removed. In the last step, all remaining units are rated as either being positive, neutral or negative. Only the negative units are counted and thereby build the score for negative cognitions. All of these steps are performed independently by multiple researchers in order to increase objectivity.[3]

Lindsey's Questionnaire

The second one is a questionnaire used in a study by Lindsey [27]. The goal of the study was not to develop a questionnaire for state reactance. State reactance only played a small role in the study [27, 34]. The questionnaire uses four items that are modified from items of Hong's Psychological Reactance Scale [19–21](see Sect. 2.3.1.4). The items are formulated as follows [27]:

1. I am uncomfortable that I am being told how to feel about bone marrow donation.
2. I do not like that I am being told how to feel about bone marrow donation.
3. It irritates me that the message told me how to feel about bone marrow donation.
4. I dislike that I am being told how to feel about bone marrow donation.

It is a unidimensional measure and does not provide distinct dimensions for anger and negative cognitions. There is only one item (item 3) that shows face-validity to address the anger component of state reactance and none that addresses negative cognitions. Also, items 2 and 4 are probably redundant. It is claimed that the items are based on Hong's Psychological Reactance Scale (HPRS), but only item 3 resembles an item from the scale because it uses the phrase "It irritates me". Still, Quick conducted a review of the scale, comparing it with the measure of Dillard and Shen [10, 34]. He concluded that the scale showed an excellent internal consistency with a Cronbach's α > 0.9,[4] but also noted the shortcoming of not distinguishing between anger, negative cognitions, and freedom threat.

Reactance Restoration Scale

Another questionnaire that addresses state reactance is the Reactance Restoration Scale (RRS) by Quick and Stephenson [33]. It consists of three statements that have

---

[3]This technique was applied in Sect. 9.3 for the validation of a state reactance questionnaire.
[4]Cronbach's α is a measure of internal consistency, which can range from 0 to 1, with 1 being of perfect internal consistency [9].

to be tailored to a specific situation. An example statement provided is: "Right now, I am—to (exercise/use sunscreen the next time I am exposed to direct sunlight for an extended period of time [greater than 15 min])" [33, p. 134]. Each statement has to be responded to by four seven-point semantic differential items:

1. motivated—unmotivated
2. determined—not determined
3. encouraged—not encouraged
4. inspired—not inspired

The questionnaire has been validated with the state reactance measure of Dillard and Shen and the trait reactance measure of HPRS (revised version) [10, 19, 33]. Significant correlations were found between the RRS and the two other measures, but the amount of correlation was rather low: The correlation of the three items of the questionnaire with state reactance (measure by Dillard and Shen [10]) were $r = 0.43$, $r = 0.14$ and $r = 0.26$ for one assessed condition and $r = 0.33$, $r = 0.12$ and $r = 0.14$ for the second assessed condition. Correlation with Hong's Psychological Reactance Scale was even lower, not exceeding $r = 0.2$ in any of the six measurement points (three items over two conditions). Also, the Reactance Restoration Scale showed mediocre correlations between $r = 0.46$ and $r = 0.54$ with all three dimensions of the Salzburger state reactance Scale [44].

Salzburger State Reactance Scale

The newest questionnaire is the Salzburger State Reactance Scale (SSRS), which is a ten-item questionnaire [23, 44]. The scale consists of the three factors experience of reactance, negative attitudes, and aggressive behavioral intentions. Experience of reactance can be regarded as a control factor that assesses a perceived threat to freedom. It includes items like "To what extent do you perceive the reaction [...] as a restriction of freedom?". The second dimension is named negative attitudes. This dimension includes items like "Do you think that [...] could also have prejudices against foreigners" and is therefore probably related to negative cognitions. The third dimension is aggressive behavioral intentions, an example item of this dimension is "Would you like to ruin [...] reputation by publishing a negative review on a relevant Internet site?". All items are answered via a five-point Likert scale [26].

The questionnaire is well validated and shows correlations with other reactance scales like the RRS [33] and HPRS [19].

2.3   Trait Reactance

When psychological reactance theory emerged, reactance was described as a psychological state and not as a personality trait. It was Schwartz [42, p. 62], who first noticed that some people were apparently more prone to becoming reactant during experiments. Schwartz conducted an experiment in which participants, who had completed personality tests before, received appeals to raise money via telephone.

Those people who showed the highest awareness regarding the consequences of a situation offered less help when the request for help was highly salient, whereas those people who were less aware of the consequences offered more help if the request was highly salient. With requests of low salience, the effect was the other way around. The people with a high awareness of consequences offered more help than the people with a low awareness of consequences [42, p. 61]. In such a scenario, request salience would correspond to freedom threat, because a highly salient request implies a stronger urge for help and thereby threats the freedom not to help of the addressee. Schwartz speculated, that persons who are very aware of consequences might have a lower threshold for experiencing state reactance [42]. Also, Brehm [2, p. 21] noted that some clinical patients showed a lower threshold of becoming reactant.

Such observations triggered some research on trait reactance. However, most of it dealt with its measurement (see Table 2.1). Apart from trait reactance measurement, some researchers tried to figure out the role of trait reactance in the formation of reactance effects. For example, Dillard and Shen [10] found that trait reactance was a significant predictor for state reactance. Additionally, Quick and Stephens found that trait reactance correlates with boomerang effects, caused by state reactance [33]. These findings suggest that there is a causal connection between trait reactance and state reactance and that trait reactance might play a role in the formation of state reactance, e.g., as a moderator variable.

2.3.1 Measurement

The observations from the listed experiments and studies [42, p. 62] [2, p. 21] made it apparent that there is a personality predisposition that determines a person's proneness of entering a reactant state.

2.3.1.1 Awareness of Consequences Scale

Already Schwartz had found a tool that might be able to predict such a predisposition. People who scored high on his Awareness of Consequences Scale [41], are more likely to experience state reactance [41]. Therefore, one could argue that the Awareness of Consequences Scale is the first tool that was able to access trait reactance. However, this line of research was not continued and assessment using the scale is rather complicated. The Awareness of Consequences scale is a collection of six short stories, describing incidents in which the protagonist faces a decision. The participants have to describe the thoughts and feelings of the protagonist. Their responses are then coded and rated by independent researchers [41].

2.3.1.2   Therapeutic Reactance Scale

The Therapeutic Reactance Scale (TRS) was developed by Dowd et al. and contains 28 items [11]. The scale measures trait reactance on two dimensions that are named behavioral reactance and verbal reactance. Behavioral reactance is assessed with 17 items, one example item for behavioral reactance is "If I am told what to do, I often do the opposite". Verbal reactance is assessed with 11 items, one example item is "In discussions, I am easily persuaded by others." [11]. Dowd et al. reported intermediate internal consistency with Cronbach's $\alpha = 0.81$ for behavioral reactance and a Cronbach's $\alpha = 0.75$ for verbal reactance. Taken as a whole, Cronbach's α was 0.84.

2.3.1.3   Fragebogen zur Messung der Psychologischen Reaktanz

Because of the lack of an adequate operationalization, Merz started to develop a quantitative self-report tool that was intended to measure a person's level of trait reactance. His "Fragebogen zur Messung der Psychologischen Reaktanz"[5] (FBMPR) was published in 1983 [29]. In his paper, Merz described the development process of the 18 items that are assessed with a six-point Likert-scale [26, 29]. Merz reported a four-factor structure that accounted for 53% of the variance with an internal consistency of $\alpha = 0.84$ and concluded that these results were satisfactory [29]. However, Herzberg challenged this conclusion and argued that the psychometric properties of the scale were insufficient [18]. He used a series of studies and subsequently deleted six of the 18 initial items and changed the format to a 4-point Likert-Scale [18, 26]. This resulted in a twelve-item questionnaire with one factor that, according to Herzberg, shows the same retest reliability than the original scale. Herzberg did not report how much variance his version of the FBMPS explained, neither did he report Cronbach's α.[6]

Meanwhile, Tucker and Byers had translated the FBMPS into English and analyzed the psychometric properties of the translated version. Their analysis suggested a two-factor solution with only 21% of explained variance [48]. Because of the low amount of explained variance and the different factor structure, they argued that the translated version needed "extensive semantic purification" and was "psychometrically unacceptable" [48, p. 814]. The data that Tucker and Byers based their analysis on was collected using a sample from midwestern USA [48]. Hong and Ostini also evaluated the psychometric properties, of the translation that Tucker and Byers created, on an Australian sample [22] and reported different results. Hong and Ostini found a four-factor structure that was similar to the four-factor structure that Merz reported [22, 29], but which explained only 44.1% of the variance in total. Like

---

[5]EN: "questionnaire for the assessment of psychological reactance".

[6]He argued that providing Cronbach's α was not necessary, because the model validity was proven for a rating-scale model. Instead of Cronbach's α, indices for individual items were provided.

Tucker and Byers, also Hong and Ostini concluded that the English version of the FBMPS was psychometrically unstable.

2.3.1.4 Hong's Psychological Reactance Scale

The English version of Merz's FBMPS was regarded as psychometrically unstable by Hong and Ostini [22] and others, because researchers either failed to replicate the four-factor structure that Merz found at all [48], or found that the four-factor structure showed poor reliablity [22]. As a consequence, Hong and Page started developing a new trait reactance questionnaire [20]. They used the English translation of Merz's scale [48] as a basis.

Within a student project, the items that Tucker used were refined according to unspecified criteria from [12, 25] to form a pool of 60 items. The students then selected a subset of 15 items from this initial pool and presented them to nine experts, who removed one of these items. The procedure resulted in a set of 14 items that were evaluated in a retest-scenario. Participants completed the questionnaire twice with an interval of two weeks in between. Hong and Page reported that the questionnaire showed a clear four-factor structure that explained 52.7% of the variance in total. Cronbach's α, a measure for internal consistency, was reported to be 0.77.

In a subsequent publication, Hong confirmed the four-factor structure of his questionnaire and reported the explained % of variance to be 55.4 with a Cronbach's α of 0.81 [21]. According to Weise, a Cronbach's α score of 0.8 to 0.9 can be considered as medium reliable [49]. Still, Hong later published a refinement of the scale, deleting three items in the process. The new version of the questionnaire showed a Cronbach's α of 0.77 [19], which is below the range of medium reliability as proposed by Weise.

Questions about the factor structure of the scale were raised by Thomas et al. [47], who tested a confirmatory model of the scale including a unifactorial solution. Thomas et al. eventually concluded that the four-factor solution would be the most likely one. However, Shen and Dillard revised the data that Thomas et al. used [47] and argued that a uni-factorial model is supported by that data [43]. Shen and Dillard also analyzed new data in a confirmatory approach by comparing different models. Also with their new data, evidence supported a second-order uni-factorial solution with the original four factors as first-order factors [43]. Furthermore, Jonason and Knowles argued that the studies which had found multiple factors via exploratory factor analysis before might have used false assumptions [24]. For example, they state that Hong et al. [19, 22], and also Tucker and Byers [48] had used oblique or oblimin rotation techniques that force factors to be orthogonal or unrelated.[7] They further suggest that factors of a construct like reactance are more likely to be related to each other, which is also supported by a correlation of r = 0.24 between factors that has been reported by Tucker and Byers [24, 48]. Brown et al. conducted another review and compared multiple models of the scale. Their conclusion was, that the

---

[7]Tucker and Byers based their work on their translation of the FBMPS [29, 48], however, these items were later used as a basis of the HPRS [20].

Table 2.1 History of Hong's Psychological Reactance Scale

1983	Development of "Fragebogen zur Messung der Psychologischen Reaktanz" (4-factor solution) [29]
1987	Translation into English (2-factor solution) [48]
1989	Further Evaluation of Merz's psychological reactance Scale (4-factor solution (unstable)) [22]
1989	Development of Hong's Psychological Reactance Scale, based on the translation by Tucker et al. [48] (4-factor solution) [20]
1992	Further Factor Analytic Validation of Hong's Psychological Reactance Scale (4-factor solution) [21]
1996	Refinement of Hong's Psychological Reactance Scale (4-factor solution) [19]
2001	Confirmatory Factor Analysis on Hong's Psychological Reactance Scale (4(1)-factor solution) [47]
2005	Analysis of the psychometric properties of Hong's Psychological Reactance Scale (1-factor) [43]
2006	Unidimensional Measure of Hong's Psychological Reactance Scale (1-factor solution) [24]
2011	Dimensionality of Hong's Psychologica Reactance Scale (1(4)-factor solution) [5]
2018	Subset of Hong's Psychologial Reactance Scale (1-factor solution) [50]

best fitting model was a bifactor model with one general factor for trait reactance and four factors for different aspects of trait reactance [5]. Based on the findings by Brown et al. [5], Yost et al. created a unidimensional version of the HPRS, that contained only eight items [50]. An overview of the development and the results regarding dimensionality is provided in Table 2.1.

2.3.2 Correlation with State Reactance

Section 2.2.3 already introduced trait reactance as a possible moderator variable for state reactance in the context of human–computer interaction. For other contexts, this assumption is supported by findings of correlations of trait reactance with state reactance. For example, Sittenthaler et al. found correlations of trait reactance, assessed with several measures, namely with the refined version of HPRS [19], with Herzbergs optimized version of Merz's FBMPR [18, 29] and with Dowds TRS [11, 44]. Also, Yost et al. found a positive correlation between trait reactance measured by HPRS (Items 1,6,9,10,11,12) and state reactance, which was measured via the Dillard and Shen method [10, 50].

2.4 Intermediate Discussion on Reactance Assessment

Being able to adequately measure reactance, whether as a personality trait or a motivational state, is vital to investigate its role for human–computer interaction. A variety of questionnaires and tools that measure state reactance and trait reactance have been introduced in Sects. 2.2.4 and 2.3.1. First, the measurements for state reactance shall be discussed.

2.4.1 State Reactance Measurement

Assessment of state reactance is possible by a variety of methods. The first method that was used to assess state reactance were behavioral observations. Scholars recorded the behavior of participants and drew conclusions about their state of reactance according to how much they engaged in behavior that is associated with reactance, such as source derogation. Such measures are quite costly in terms of human labor and are also prone to misjudgment. Furthermore, behavioral observations as reactance measures were done out of necessity in the past, because there was no other operationalization of state reactance. In fact, measuring reactance was thought to be impossible, Brehm and Brehm [4, p. 37] wrote:

> We cannot measure reactance directly, but hypothesizing its existence allows us to predict a variety of behavioral effects.

Dillard and Shen Method

When Dillard and Shen published their method of measuring state reactance as an operationalization of anger and negative cognitions, it was finally possible to get a nominal value of state reactance with relatively low effort. Besides introducing their measurement technique, they were also able to show that psychological reactance is probably consisting of anger and negative cognitions that are entangled in a way, that their individual effect on behavior is unpredictable [10]. Dillard's and Shen's method has been validated several times [34, 36] and its operationalization of anger via a short questionnaire, with items that are formulated very generally and negative cognitions as a thought-listing task, it can be applied in a variety of contexts, including human–computer interaction. The downside of the method is that it still requires human intervention for measuring negative cognitions. The thought-listing task takes quite long and requires participants to closely interact with the experimenter. This situation increases the risk of social desirability effects.[8]

Lindsey's Questionnaire

In the same year, Lindsey published a study that included a short questionnaire which could also assess state reactance [27], even though her questionnaire did

---

[8]A social desirability effect is when a participant reports or replies in a way that he or she thinks is socially desirable, instead of his or her true opinion [13].

not allow for distinguishing between the components of state reactance. Lindsey's questionnaire consists of four items only, two of which are probably redundant. This might also be one reason for its excellent internal consistency. Sittenthaler et al. found that Lindsey's questionnaire correlates highly (r = 0.75) with the dimension *experience of reactance*, that is part of their Salzburger state reactance Scale. Also, the dimensions *aggressive behavioral intentions* and *negative attitudes* showed a mediocre correlation with the scale (r = 0.41 and r = 0.35). The fact that Lindsey's questionnaire is short and consists of only four items makes it attractive for the use of state reactance assessment in the context of human–computer interaction research and for online studies.

Reactance Restoration Scale

The Reactance Restoration Scale is well validated [32, 33, 44] and measures boomerang effects, which are consequences of state reactance. The method itself is a bit complicated because there are three statements as cloze texts, that have to be answered via four semantic differential items. The statements are formulated in a way that they address behavior and motivation of the subjects. Also, the items are directly addressing health-related behavior, so it is questionable if reformulating them to the context of human–computer interaction would result in a measure that is still valid.

Salzburger State Reactance Scale

The Salzburger state reactance Scale has been shown to correlate with most of the other measures of psychological reactance [44], which is a strong argument for its validity. But the formulation of its items is quite specific, e.g., "Do you think that this landlord could also have prejudices against foreigners?" [44] and it is questionable if it could be adapted to be useful for human–computer interaction research.

2.4.2   Trait Reactance Assessment

Three different measurement tools have been identified in the literature search. Their basic properties and their usefulness for human–computer interaction research are discussed in the following paragraphs.

Therapeutic Reactance Scale

The Therapeutic Reactance Scale by Quick et al. is the longest of the three trait reactance questionnaires that were found in the literature. It contains 28 items that are used to assess trait reactance on the two factors *behavioral reactance* and *verbal reactance*. Even though the Therapeutic Reactance Scale is quite long, it can in principle be used for human–computer interaction research, both in laboratory experiments and in online studies, because its items do not need to be tailored to a specific situation.

Fragebogen zur Messung der Psychologischen Reaktanz

The Fragebogen zur Messung der Psychologischen Reaktanz [29] has been the first questionnaire that was aimed at measuring trait reactance. However, its original form and also its translations into English were deemed to be psychometrically unstable [18, 48]. While the English version was used as a basis for Hong's Psychological Reactance Scale, the original version was revised by Herzberg, who removed several items and interpreted it with a 1-factor solution [18]. Merz's original scale, as well as Herzberg's revised version, are useful for assessing trait reactance for human–computer interaction research because their items address general attitudes of the participants and do not need to be tailored to a specific situation.

Hong's Psychological Reactance Scale

A literature search has revealed a large body of work that has been conducted on Hong's Psychological Reactance Scale [19, 20] and its predecessor, the Fragebogen zur Messung der Psychologischen Reaktanz [29], over a period of 35 years. Scholars seem to be unsure how to interpret the results of Hong's Psychological Reactance Scale, mostly because of its unclear factor structure. Most scholars seem to argue for either a 4-factor or a 1-factor solution.[9] The most recent research even argues for a removal of several items, in favor of a 1-factor solution. Despite the uncertainty about its factor structure, Hong's Psychological Reactance Scale can, in principle be used in the context of human–computer interaction, because, like the other two questionnaires, its items are not formulated to address the current situation, but rather general attitudes of the subject.

2.5 Conclusion on Reactance Assessment

All three questionnaires for assessing trait reactance that have been found in the literature can be used in human–computer interaction research, in principle. This is because they all use self-report items, which can easily be applied to laboratory experiments and online studies. Another advantage that all three questionnaires share is that their items are formulated to address the general attitudes of the participants and therefore do not require to be adapted to the context of the study they are used in. Still, there are some differences between the questionnaires that reduce their usefulness in practice. The Therapeutic Reactance Scale is quite long and would take the participants considerable time to complete. Furthermore, it has a 2-dimensional structure with dimensions that address trait reactance from a verbal and a behavioral point of view. In contrast to state reactance, there is no formal conceptualization that would explain why trait reactance should consist of multiple factors. This makes interpretation of the results quite difficult. Keeping this in mind, and following the consideration that a practical measurement tool for trait reactance should be as short

---

[9]See Table 2.1 for an overview of the research that has been conducted and the resulting factor structure.

as possible,[10] it is concluded that two trait reactance questionnaires are to be used further during this work. These questionnaires are the HPRS, using the 1-factor solution by Yost [19, 20, 50] and the FBMPR in the revised version by Herzberg [18].

Several methods have been introduced that can measure either state reactance or trait reactance in different contexts. In order to be useful for the current book, a state reactance measure should be able to assess state reactance in the context of human–computer interaction research. In the case of a questionnaire, this requires the items to be formulated in a way that they can address the user's experience while interacting with a technical system. The Salzburger state reactance Scale [23, 44] and the Reactance Restoration Scale [33] are well-constructed questionnaires that are both validated, but their items are formulated in a quite specific manner. This is to such an extent, that both are probably unsuitable for the context of human–computer interaction.

The questionnaire of Lindsey, on the other hand, is formulated in a way that it can easily be reformulated to fit other contexts, such as human–computer interaction. Additionally, it is quite short, straightforward and can be answered quickly. This characteristic would make it ideal for the use in online studies and laboratory experiments, alongside other questionnaires. On the downside, Lindsey's questionnaire is only unidimensional and does not distinguish between the components of state reactance. This results in the measures of Lindsey's questionnaire being of little use for thorough assessment of the role of state reactance in human–computer interaction.

Out of the reviewed methods of state reactance measurements, the method of Dillard and Shen is the most promising one. The anger scale is very short and the method can distinguish between the components of state reactance. Additionally, the method can be easily applied to different domains without the need to adapt it. But, even though it is certainly useful for research on state reactance, its thought-listing task is too costly to use for frequent measurement of state reactance in the laboratory and hardly usable for online studies.

In sum, the review of available state reactance measures did not result in an adequate measurement tool to easily assess state reactance in the context of human–computer interaction and that can be used for laboratory experiments and online studies. For this reason, a new measurement tool shall be developed in the form of a self-report questionnaire. Its development is described in Chap. 8.

References

1. Bortz, J., Dring, N.: Forschungsmethoden und Evaluation für Statistik für Human- und Sozial-wissenschaftler, 4th edn. Springer (2006)
2. Brehm, S.S.: The Apllication of Social Psychology to Clinical Practice. Hemisphere Publishing Corporation (1976)
3. Brehm, J.W.: A Theory of Psychological Reactance. Academic Press, New York (1966)

---

[10]Questionnaires should be as short as possible to be easily included in the test battery of laboratory experiments and online studies.

4. Brehm, S.S., Brehm, J.W.: Psychological Reactance: A Theory of Freedom and Control. Academic Press, New York (1981)
5. Brown, A.R., Finney, S.J., France, M.K.: Using the bifactor model to assess the dimensionality of the hong psychological reactance scale. Educ. Psychol. Measurement **71**(1), 170–185 (2011)
6. Burgoon, M., Alvaro, E.M., Broneck, K., Miller, C., Grandpre, J.R., Hall, J.R., Frank, C.A.: Using interactive media tools to test substance abuse prevention messages. In: Crano, W.D., Burgoon, M. (Eds.), Mass Media and Drug Prevention: Classic and Contemporary Theories and Research, pp. 67–87. Lawrence Erlbaum Associates Publishers, Mahwah, NJ, US
7. Byrne, B.M.: Structural Equation Modeling with LISREL, PRELIS, and SIMPLIS: Basic Concepts, Applications, and Programmings. Psychology Press, Tylor & Francis Group (1998)
8. Carpinella, C.M., Wyman, A.B., Perez, M.A., Stroessner, S.J.: The robotic social attributes scale (rosas): development and validation, pp. 254–262 (2017). https://doi.org/10.1145/2909824. 3020208
9. Cronbach, L.J.: Coefficient alpha and the internal structure of tests. Psychometrika **16**(3), 297–334 (1951). https://doi.org/10.1007/BF02310555
10. Dillard, J.P., Shen, L.: On the nature of reactance and its role in persuasive health communication. Commun. Monographs **72**(2), 144–168 (2005). https://doi.org/10.1080/03637750500111815
11. Dowd, E.T., Milne, C.R., Wise, S.L.: The therapeutic reactance scale: A measure of psychological reactance. J. Couns. Dev. **69**(6), 541–545 (1991). https://doi.org/10.1002/j.1556-6676.1991.tb02638.x
12. Edwards, A.: Techniques of Attitude Scale Construction. Century psychology series. Irvington Publishers (1983). URL https://books.google.de/books?id=VAowWzh5r08C
13. Edwards, A.L.: The relationship between the judged desirability of a trait and the probability that the trait will be endorsed. J. Appl. Psychol. **37**(2), 90 (1953)
14. Ehrenbrink, P., Osman, S., Möller, S.: Google Now is for the Extraverted, Cortana for the introverted: Investigating the influence of personality on IPA preference. In: Proceedings of the 29th Australian Conference on Human-Computer Interaction, pp. 1–9. ACM, New York, NY (2017). https://doi.org/10.1145/3152771.3152799. Electronic, online
15. Field, A.: Discovering Statistics Using IBM SPSS Statistics, vol. 4. SAGE Publications Ltd (2013)
16. Ghazali, A.S., Ham, J., Barakova, E., Markopoulos, P.: The influence of social cues in persuasive social robots on psychological reactance and compliance. Comput. Human Behav. **87**, 58–65 (2018). https://doi.org/10.1016/j.chb.2018.05.016. URL http://www.sciencedirect.com/science/article/pii/S0747563218302425
17. Grandpre, J., Alvaro, E.M., Burgoon, M., Miller, C.H., Hall, J.R.: Adolescent reactance and anti-smoking campaigns: a theoretical approach. Health Commun. **3**, 349–366 (2003)
18. Herzberg, P.Y.: Zur psychometrischen Optimierung einer Reaktanzskala mittels klassischer IRT-basierter Analysemethoden. Diagnostica **48**(4), 163–171 (2002). https://doi.org/10.1026//0012-1924.48.4.163
19. Hong, S.M., Faedda, S.: Refinement of the hong psychological reactance scale. Educ. Psychol. Measurement **56**(1), 173–182 (1996). https://doi.org/10.1177/0013164496056001014
20. Hong, S.M., Page, S.: A psychological reactance scale: development, factor structure and reliability. Psychol. Reports **64**(3_suppl), 1323–1326 (1989). https://doi.org/10.2466/pr0.1989.64.3c.1323
21. Hong, S.M.: Hong's psychological reactance scale: a further factor analytic validation. Psychol. Reports **70**(2), 512–514 (1992). https://doi.org/10.2466/pr0.1992.70.2.512
22. Hong, S.M., Ostini, R.: Further evaluation of merz's psychological reactance scale. Psychol. Reports **64**(3), 707–710 (1989)
23. Jonas, E., Graupmann, V., Kayser, D.N., Zanna, M., Traut-Mattausch, E., Frey, D.: Culture, self, and the emergence of reactance: is there a universal freedom? J. Exp. Soc. Psychol. **45**(5), 1068–1080 (2009)
24. Jonason, P.K., Knowles, H.M.: A unidimensional measure of hong's psychological reactance scale. Psychol. Reports **98**(2), 569–579 (2006). https://doi.org/10.2466/pr0.98.2.569-579. PMID: 16796116

25. Kidder, L., Judd, C., Smith, E., for the Psychological Study of Social Issues, S.: Research methods in social relations. Holt, Rinehart and Winston (1986). URL https://books.google.de/books?id=i_V9AAAAIAAJ
26. Likert, R.: A technique for the measurement of attitudes. Archi, Psychol (1932)
27. Lindsey, L.L.M.: Anticipated guilt as behavioral motivation an examination of appeals to help unknown others through bone marrow donation. Human Commun. Res. **31**(4), 453–481 (2005)
28. Mayer, R.E., Sobko, K., Mautone, P.D.: Social cues in multimedia learning: role of speaker's voice. J. Educ. Psychol. **95**(2), 419–425 (2003). https://doi.org/10.1037/0022-0663.95.2.419
29. Merz, J.: Fragebogen zur messung der psychologischen reaktanz. Diagnostica **29**(1), 75–82 (1983)
30. Nass, C., Moon, Y.: Machines and mindlessness: social responses to computers. J. Social Issues **56**(1), 81–103 (2000). https://doi.org/10.1111/0022-4537.00153. URL https://spssi.onlinelibrary.wiley.com/doi/abs/10.1111/0022-4537.00153
31. Pervin, L.A., John, O.P.: Personality Theory and Research, 8th edition. John Wiley & Sons, Inc. (2001)
32. Quick, B.L., Stephenson, M.T.: Examining the role of trait reactance and sensation seeking on perceived threat, state reactance, and reactance restoration. Human Commun. Res. **34**(3), 448–476 (2008). https://doi.org/10.1111/j.1468-2958.2008.00328.x
33. Quick, B.L., Stephenson, M.T.: The reactance restoration scale (rrs): a measure of direct and indirect restoration. Commun. Res. Reports **24**(2), 131–138 (2007). https://doi.org/10.1080/08824090701304840
34. Quick, B.L.: What is the best measure of psychological reactance? an empirical test of two measures. Health Commun. **27**(1), 1–9 (2012). https://doi.org/10.1080/10410236.2011.567446. PMID: 21714621
35. Quick, B.L., Scott, A.M., Ledbetter, A.M.: A close examination of trait reactance and issue involvement as moderators of psychological reactance theory. J. Health Commun. **16**(6), 660–679 (2011). https://doi.org/10.1080/10810730.2011.551989
36. Rains, S.A.: The nature of psychological reactance revisited: a meta-analytic review. Human Commun. Res. **39**(1), 47–73 (2013). https://doi.org/10.1111/j.1468-2958.2012.01443.x
37. Rains, S.A., Turner, M.M.: Psychological reactance and persuasive health communication: a test and extension of the intertwined model. Human Commun. Res. **33**(2), 241–269 (2007). https://doi.org/10.1111/j.1468-2958.2007.00298.x
38. Roubroeks, M., Ham, J., Midden, C.: The dominant robot: Threatening robots cause psychological reactance, especially when they have incongruent goals. In: Ploug, T., Hasle, P., Oinas-Kukkonen, H. (eds.) Persuasive Technology, pp. 174–184. Springer, Berlin, Heidelberg (2010)
39. Roubroeks, M., Ham, J., Midden, C.: When artificial social agents try to persuade people: the role of social agency on the occurrence of psychological reactance. Int. J. Soc. Robot. **3**(2), 155–165 (2011). https://doi.org/10.1007/s12369-010-0088-1
40. Roubroeks, M., Midden, C., Ham, J.: Does it make a difference who tells you what to do ? exploring the effect of social agency on psychological reactance. In: Proceedings of the 4th International Conference on Persuasive Technology, Persuasive '09, pp. 15:1–15:6. ACM, New York, NY, USA (2009). https://doi.org/10.1145/1541948.1541970
41. Schwartz, S.H.: Awareness of consequences and the influence of moral norms on interpersonal behavior. Sociometry, pp. 355–369 (1968)
42. Schwartz, S.H.: Awareness of interpersonal consequences, responsibility denial, and volunteering. J. Personal. Soc. Psychol. **30**(1), 57 (1974)
43. Shen, L., Dillard, J.P.: Psychometric properties of the hong psychological reactance scale. J. Pers. Assess. **85**(1), 74–81 (2010)
44. Sittenthaler, S., Traut-Mattausch, E., Steindl, C., Jonas, E.: Salzburger state reactance scale (ssr scale): Validation of a scale measuring state reactance. Zeitschrift für Psychologie **223**, 257–266 (2015). https://doi.org/10.1027/2151-2604/a000227
45. Smith, M.J.: The effects of threats to attitudinal freedom as a function of message quality and initial receiver attitude. Commun. Monogr. **44**(3), 196–206 (1977). https://doi.org/10.1080/03637757709390131

46. Steindl, C., Jonas, E., Sittenthaler, S., Traut-Mattausch, E., Greenberg, J.: Understanding psychological reactance. Zeitschrift für Psychologie **223**, 205–214 (2015). https://doi.org/10.1027/2151-2604/a000222
47. Thomas, A., Donnell, A.J., Buboltz Jr., W.C.: The hong psychological reactance scale: A confirmatory factor analysis. Measurement Eval. Couns. Dev. **34**(1), 2 (2001)
48. Tucker, R.K., Byers, P.Y.: Factorial validity of merz's psychological reactance scale. Psychol. Reports **61**(3), 811–815 (1987). https://doi.org/10.2466/pr0.1987.61.3.811
49. Weise, G.: Psychologische Leistungstests: ein Handbuch für Studium und Praxis. Verlag für Psychologie Hogrefe, Psychologische Leistungstests (1975)
50. Yost, A.B., Behrend, T.S., Howardson, G., Badger Darrow, J., Jensen, J.M.: Reactance to electronic surveillance: a test of antecedents and outcomes. J. Bus. Psychol. (2018). https://doi.org/10.1007/s10869-018-9532-2

Part II
Relevance of Psychological Reactance in Human–Computer Interaction

Chapter 3
Is Psychological Reactance Relevant for Human–Computer Interaction? If Yes, in Which Context?

3.1 Research Question

Research Question 1 *State reactance is a consequence of a perceived loss of freedom or control. Also, state reactance is a social phenomenon. Brehm and Brehm argued, that a threat to freedom can cause reactance only if the person really knows that he or she can exercise this freedom [1, p.12]. Technical devices and services are deterministic machines that follow rules. Users of such machines consciously know that. Therefore, it is reasonable to assume that a freedom threat, which is posed by a machine, is regarded as given and does not trigger state reactance. On the other hand, it was shown that humans who are interacting with computers can just ignore such considerations and apply schemes that they know from human–human interaction [2]. Research Question 1 therefore asks: Is psychological reactance relevant for human-computer interaction?*

Research Questions 1.1 and 1.2 follow from Research Question 1:

Research Question 1.1 *Persons can experience state reactance when they perceive a threat of freedom. It is currently unknown in what situations personal freedoms are perceived as being under threat by technical devices or services. Provided that state reactance can be triggered by a machine, what are the situations in which state reactance is usually triggered?*

Research Question 1.2 *Section 2.2.1 already provided an overview of the possible consequences that follow if a person experiences state reactance. Provided that state reactance can be triggered by technical devices or services, what are the consequences thereof?*

© Springer Nature Switzerland AG 2020
P. Ehrenbrink, *The Role of Psychological Reactance in Human–Computer Interaction*, T-Labs Series in Telecommunication Services,
https://doi.org/10.1007/978-3-030-30310-5_3

3.2 Methods

Research Question 1 is a very broad one. To get an overview of the phenomenon of psychological reactance in the wide field of human–computer interaction, a combination of two strategies was employed. The first one was a literature search. In a systematic literature search, papers that discuss the topic of psychological reactance in connection with technical devices and services were collected. Afterward, the included studies were analyzed in terms of whether state reactance was triggered in users or whether trait reactance had any effect on human–computer interaction.

In the next step, Research Question 1.1 was answered by a combination of literature review and a qualitative survey. The literature that has already been analyzed for Research Question 1 were clustered according to the situation in which reactance is investigated. Additionally, a qualitative survey among usability experts was conducted to identify situations where state reactance is triggered during everyday interaction with technical devices. Also, a laboratory experiment was conducted, in order to find out if technical devices or services can pose a freedom threat.

Finally, to answer Research Question 1.2, the identified literature, the results of the qualitative survey and the results of the laboratory experiment are analyzed according to possible consequences of state reactance.

References

1. Brehm, S.S., Brehm, J.W.: Psychological Reactance: A Theory of Freedom and Control. Academic Press, New York (1981)
2. Nass, C., Moon, Y.: Machines and mindlessness: social responses to computers. J. Social Issues **56**(1), 81–103 (2000). https://doi.org/10.1111/0022-4537.00153, https://spssi.onlinelibrary.wiley.com/doi/abs/10.1111/0022-4537.00153

Chapter 4
Literature Search—Reactance in Literature

4.1 Introduction

Chapter 1 already introduced the concept of psychological reactance and its two-fold conceptualization as being a motivational state and a personality trait. Since the introduction of reactance theory by Brehm [2], research on the topic has been conducted mostly in the fields of marketing, psychology, communication and health [1]. In contrast, little research has been dedicated to psychological reactance in the domain of human–computer interaction. It was argued in Chapter , that the rise of intelligent systems can lead to the higher importance of phenomena which are transferred from human–human interaction to human–computer interaction. Following this argument, the work that is described in this thesis was conducted. While the mechanism that triggers state reactance is known to be a perceived loss of freedom or control, typical situations, in which such a threat is perceived are still a matter of speculation. The same goes for the question if state reactance can be a consequence of interaction with technical devices, at all. Therefore, as a first step of exploring the role of psychological reactance in human–computer interaction, situations that can lead to state reactance shall be identified.

4.2 Results

The papers that were identified in the literature search were analyzed in order to answer Research Questions 1 and 1.1. The results are presented in the upcoming sections.

© Springer Nature Switzerland AG 2020
P. Ehrenbrink, *The Role of Psychological Reactance in Human–Computer Interaction*, T-Labs Series in Telecommunication Services,
https://doi.org/10.1007/978-3-030-30310-5_4

4.2.1   Research Question 1

In order to be able to answer Research Question 1, all papers that were identified in the literature search were reviewed according to whether the described studies were able to gather evidence that either state reactance was induced because of the participants' interaction with technical devices or services, or that trait reactance had an effect on interaction. The results are shown in Table 4.1.

4.2.2   Research Question 1.1

All identified papers were read and categorized independently by two researchers in a double-blind coding procedure. Afterward, the researchers conducted a workshop and consolidated the categories that they had built. The categories that the two researchers had built were mostly overlapping, but one group (Reduced Freedom of Choice) was split in two (Reduced Freedom of Choice and Forced Response). The resulting groups were then named according to the identified situation that reactance played a role in. Seven categories of situations in which reactance was playing a role in the context of human–computer interaction were identified.

Persuasive Attempts

I total, the literature search ended up with five papers that dealt with the concept of reactance in connection with persuasive attempts. The first paper introduced the idea of a smart key holder that was supposed to persuade its users into using their

Table 4.1 Relevance of state reactance and trait reactance for human–computer interaction. Cases where reactance was not measured, but used as an explanation for observed effects are marked as indirect

	Evidence that state reactance was triggerd	Evidence that trait reactance played a role
Laschke et al. [8]	Not considered	Not considered
Zheng et al. [18]	Indirect	Not considered
Murray and Häubl [12]	Indirect	Not considered
Roubroeks et al. [15]	Yes	Not considered
Roubroeks et al. [13, 14]	Yes	Not considered
Lee and Lee [9]	Indirect	Not considered
White et al. [17]	Indirect	Not considered
Liu et al. [11]	Indirect	Not considered
Li and Meeds [10]	Not considered	Yes
Stieger et al. [16]	Indirect	Not considered
Edwards [4]	Indirect	Not considered

bicycle instead of their car. It did this by dropping the bicycle key. The user had to pick up the key and, while doing so, might decide on using the bicycle. It was postulated, that the behavior of the smart key holder would create friction which might result in the user getting reactant. The authors argue that adding ironic or naive features to the interaction might reduce state reactance during interaction [8] but did not test that hypothesis. In another paper, Liu et al. investigated the influence of the factors task/agent-orientation and persuasion during interaction with an artificial social agent that served as an assistant for a task that the participants had to perform. The social agent gave advice via either an agent-centric or a task-centric formulated text. Additionally, it could use strong persuasive (including an avatar, argumentation, and words like "certain", "best") and non-persuasive language to assist the participants in their task. Liu et al. found that participants who were in the agent-oriented group performed significantly lower on their task than participants who were in the task-oriented group. Also, participants were able to simply click a "Do as agent sais"-button and thereby perform flawlessly without much mental effort, some still choose to disagree with the agent. This effect was stronger in the persuasive condition. Liu et al. argued that this behavior might be due to psychological reactance [11]. A series of papers was published by Roubroeks et al. and also included a persuasive agent that varied in its amount of controlling language and its social agency [13–15]. They conducted an experiment with a three by three conditional design with three levels of controlling language (none, low, high) and three levels of social agency (none, a picture of a robotic cat, video of a robotic cat). The agent was an assistant system that gave advice on how to solve a task via text output. As a result, they found that the level of state reactance increased with the level of controlling language, high controlling language resulted in the highest level of state reactance. For social agency, state reactance followed the same trend but produced a less clear effect [13, 14]. The last paper of the series investigated the amount of restoration behavior [1] for congruent and incongruent goals of the persuasive agent. The results show that there is more restoration behavior in the incongruent condition when the agent gives highly threatening (strong controlling language) advice [15].

Personalization and Use of Personal Data

Two of the identified papers investigated reactance in connection with the use of personal data. The first paper, written by White et al. compared personalized email in a two by two conditional experiment. One of the two factors was the distinctiveness of personalization. There were emails that used only the name of the person for personalization (lowly distinctive) and emails that included more personal information, such as personal preferences or hobbies (highly distinctive). The other factor was justification. White et al. included emails where the personal information that was used was justified for the type of offer that was provided in the emails. Other emails used personal information that did not have a connection to the type of offer provided in the emails [17]. They found that a high level of personalization can cause state reactance

---

[1] Usually a consequence of state reactance. An attempt to restore the lost freedom of choice, e.g., by doing the opposite of what was intended by the persuasive agent.

but they also observed that this effect can be avoided by justification. They conclude that improper use of personalization causes state reactance because the users feel identifiable but an explanation of why their personal information was used, or where the information came from in the first place, can minimize this effect [17].

Lee and Lee conducted an experiment where they compared two types of only recommendation services. One service used personal information, such as registration data, browsing history, and transaction data, the other service used only the registration data [9]. Lee and Lee observed that the intention to use the service again was lower for the highly personalized service, even though the perceived usefulness was equal for both services [9]. They argued that this observation can be explained by state reactance that was triggered by the use of personal data [9].

Reduced Freedom of Choice

Even though a personal freedom or freedom of choice is the main mechanism that causes state reactance [2, 3], only one paper could be identified that explicitly dealt with reactance effects regarding reduced freedom of choice in the domain of human–computer interaction. Murray and Häubl conducted an experiment in which the participants could either choose between two interfaces, or were assigned to one interface without having the freedom of choice [12]. Afterward, they had to use a third interface. In a pre-study, the third interface was rated as being superior in ease of use and task completion time. After using the third interface, all participants were asked to choose their preferred among all interfaces. Results showed that those participants who were able to choose an interface in the first part of the experiment were more likely to prefer that one over the other interfaces, compared to those subjects, who were assigned to one interface. The interpretation of Murray and Häubl is, that assigning participants to an interface, instead of letting them choose one, resulted in a freedom threat, which then induced state reactance [12].

Forced Response

Stieger et al. conducted a study with online questionnaires. They implemented it in a way that users could neither skip items or whole questionnaires. If the users tried to skip (or forgot to reply to) at least one item of a page, they could not proceed to the next page of the questionnaire until they finished the missing items. If users tried to skip a complete questionnaire, they were presented with a website that asked them to return and complete the questionnaire that they tried to skip [16]. The researchers then compared drop-out rates of users who were aware of the restrictive implementation of the questionnaire and those users who were not aware of the restrictions (users who did not try to skip items or questionnaires). The results show that drop-out rates were higher for those users who knew about the restrictions, because they attempted to skip an item or a questionnaire, compared to those users who did not know about the restrictions. The interpretation of Stieger et al. is, that the enforcement of responses triggered (state) reactance, what in turn caused the higher drop-out rate [16].

Obstructions

Obstructions are features of functions of a technical system that prevent or disrupt users from accomplishing their task. Edwards et al. investigated effects of pop-up adds acting as obstructions while browsing the internet. In their experiment, they used two conditions. In the task-oriented condition, participants were browsing websites in the search of information. In the other condition, participants were not looking for specific information. Edwards et al. observed, that pop-up adds caused more avoidance behavior when participants were working on some task. Their interpretation was that the pop-up adds were regarded as more intrusive in such a situation and therefore cause reactance effects [4].

High Effort or Costs

Li and Meeds published an article about online advertisement on websites [10]. They noticed that persons with high trait reactance show more add avoidance than persons with low trait reactance when the add is shown only once. When the add is shown repeatedly, lowly reactant personalities show no significant change in add avoidance, while highly reactant personalities decrease their add avoidance [10]. The interpretation of Li and Meeds is that add avoidance requires mental resources and that highly reactant personalities regard this as an intrusion. This is in line with the perceived add intrusiveness which was rated high for repeated adds by highly reactant personalities, while lowly reactant personalities rated it lower.

Scarceness

Scarceness is an oddball in reactance research since scarceness is often employed intentionally in marketing to induce state reactance effects. If a product is seen as scarce, persons can perceive a threat to their freedom of owning that product. In order to secure their freedom, they will then buy it [18]. Zheng et al. conducted a series of experiments in which they investigated different strategies of scarceness marketing, including scarce time and scarce quantity. They found that scarceness advertising in online platforms can really influence users to buy more products. This includes limiting the total number of available pieces and limiting the time period in which a special offer is available.

4.2.3   Research Question 1.2

In order to gather data for answering Research Question 1.2, all identified papers were analyzed in terms of whether empirical data about consequences of state reactance had been collected. This includes direct, consequences for interaction with the system, but also intended consequences for the interaction. The results are shown in Table 4.2. The listed consequences include interaction avoidance or ignoring of messages and boomerang effects. One study also found decreased acceptance of the technical system that triggered state reactance. Zheng et al. measured an increase in

Table 4.2 List of consequences of state reactance that were reported in the identified papers

	Reported consequences
Laschke et al. [8]	No
Zheng et al. [18]	Increase unplanned buying
Murray and Häubl [12]	Decreased acceptance
Roubroeks et al. [15]	Ignoring, boomerang effects
Roubroeks et al. [13, 14]	Boomerang intentions
Lee and Lee [9]	Avoidance
White et al. [17]	Less intention to respond
Liu et al. [11]	Boomerang effects
Li and Meeds [10]	No
Stieger et al. [16]	Participation stop
Edwards [4]	Avoidance

unplanned buying, what was a positive outcome for their online shopping system, rather than a negative consequence. Some studies only observed an intention for their participants to change their interaction with the system, e.g., they wanted to do the opposite of what the persuasive system intended them to do, but no actual behavior change was observed [13, 14, 17].

References

1. Debora, Dhanya A., Pricilda Jaidev, U.: Consumer reactance: a review of research methodologies. Int. J. Pure Appl. Math. **118**(18), 4449–4467 (2018)
2. Brehm, J.W.: A Theory of Psychological Reactance. Academic Press, New York (1966)
3. Brehm, S.S., Brehm, J.W.: Psychological reactance: a theory of freedom and control. Academic Press, New York (1981)
4. Edwards, S.M., Li, H., Lee, J.H.: Forced exposure and psychological reactance: antecedents and consequences of the perceived intrusiveness of pop-up ads. J. Advert. **31**(3), 83–95 (2002). https://doi.org/10.1080/00913367.2002.10673678
5. Ehrenbrink, P., Gong, X.G., Möller, S.: Implications of different feedback types on error perception and psychological reactance. In: Proceedings of the 28th Australian Conference on Computer-Human Interaction, OzCHI '16, pp. 358–362. ACM, New York, NY, USA (2016). https://doi.org/10.1145/3010915.3010994
6. Ehrenbrink, P., Hillmann, S., Weiss, B., Möller, S.: Psychological reactance in HCI: a method towards improving acceptance of devices and services. In: Proceedings of the 28th Australian Conference on Computer-Human Interaction, OzCHI '16, pp. 478–482. ACM, New York, NY, USA (2016). https://doi.org/10.1145/3010915.3010978
7. Ehrenbrink, P., Prezenski, S.: Causes of psychological reactance in human-computer interaction—a literature review and survey. In: Proceedings of the 35th European Conference on Cognitive Ergonomics (ECCE), pp. 1–8. ACM, NY, USA (2017). https://doi.org/10.1145/3121283.3121304. Electronic

8. Laschke, M., Diefenbach, S., Schneider, T., Hassenzahl, M.: Keymoment: Initiating behavior change through friendly friction. In: Proceedings of the 8th Nordic Conference on Human-Computer Interaction: Fun, Fast, Foundational, pp. 853–858. ACM (2014)

9. Lee, G., Lee, W.J.: Psychological reactance to online recommendation services. Inf. Manage. **46**(8), 448–452 (2009). https://doi.org/10.1016/j.im.2009.07.005

10. Li, C., Meeds, R.: Factors affecting information processing of internet advertisements: A test on exposure conditions, psycholgical reactance, and advertising frequency. American Academy of Advertising. Conference of Proceedings (Online), pp. 93–101. American Academy of Advertising, Austin (2007)

11. Liu, S., Helfenstein, S., Wahlstedt, A.: Social psychology of persuasion applied to human agent interaction. Human Technol. **4**(2), 123–143 (2008)

12. Murray, K.B., Häubl, G.: Freedom of choice, ease of use, and the formation of interface preferences. MIS Quart. **35**(4), 955–976 (2011)

13. Roubroeks, M., Ham, J., Midden, C.: When artificial social agents try to persuade people: The role of social agency on the occurrence of psychological reactance. Int. J. Soc. Robot. **3**(2), 155–165 (2011). https://doi.org/10.1007/s12369-010-0088-1

14. Roubroeks, M., Midden, C., Ham, J.: Does it make a difference who tells you what to do ? exploring the effect of social agency on psychological reactance. In: Proceedings of the 4th International Conference on Persuasive Technology, Persuasive '09, pp. 15:1–15:6. ACM, New York, NY, USA (2009). https://doi.org/10.1145/1541948.1541970

15. Roubroeks, M., Ham, J., Midden, C.: The dominant robot: threatening robots cause psychological reactance, especially when they have incongruent goals. In: Ploug, T., Hasle, P., Oinas-Kukkonen, H. (eds.) Persuasive Technology, pp. 174–184. Springer, Berlin, Heidelberg (2010)

16. Stieger, S., Reips, U.D., Voracek, M.: Forced-response in online surveys: bias from reactance and an increase in sex-specific dropout. J. Am. Soc. Inf. Sci. Technol. **58**(11), 1653–1660 (2007). https://doi.org/10.1002/asi.v58:11

17. White, T.B., Zahay, D.L., Thorbjørnsen, H., Shavitt, S.: Getting too personal: reactance to highly personalized email solicitations. Market. Lett. **19**(1), 39–50 (2008)

18. Zheng, X., Liu, N., Zhao, L.: A study of the effectiveness of online scarce promotion-based on the comparison of planned buying and unplanned buying. In: Proceedings of the 12th Wuhan International Conference on E-Business, pp. 247–257 (2013)

Chapter 5
Expert Survey—Triggers for State Reactance

5.1 Introduction

The literature search already identified seven different situations in which reactance might occur in the context of human–computer interaction. In order to augment that list of situations, a qualitative study among "experts" in the field of human–computer interaction was conducted. The survey was performed in written form and collected reports about situations that triggered reactance while interacting with technology.

5.2 Methods

The survey was conducted in the context of the course "Usability Engineering" at Technische Universität Berlin.

5.2.1 Participants

21 students took part in the survey. They were all taking part in the lecture "Usability Engineering" at Technische Universität Berlin. They could not be regarded as experts with hands-on-experience in the field of human–computer interaction evaluation, but they had basic knowledge about the field and are therefore called "experts" in the following. Neither gender nor age of the participants was assessed.

© Springer Nature Switzerland AG 2020
P. Ehrenbrink, *The Role of Psychological Reactance in Human–Computer Interaction*, T-Labs Series in Telecommunication Services,
https://doi.org/10.1007/978-3-030-30310-5_5

5.2.2   Procedure

All participants received an introduction to the concept of psychological reactance. The oral presentation included a detailed description of the construct of reactance, as well as two examples of situations where persons experience reactance. The two situations were:

- A person holding an important presentation. The presentation is then interrupted by an update of the operating system. The person gets angry and starts to blame the operating system.
- Only one piece of a product is left in a store, it is certain that there will be no replenishment in the near future. Therefore, a person buys the product to avoid a situation in which he or she might no longer be able to do that.

The two situations were chosen because they cover very different settings and consequences. This was intended to illustrate the range of situations in which state reactance might play a role. At a later point, all participants were asked to write down and describe a situation in which they experienced reactance while interacting with technology.

5.3   Results

All students completed the survey, but one response was not taken into account because it did not contain any description or hint to a reactance effect.

5.3.1   Research Question 1.1

In order to be used for answering Research Question 1.1, all responses were analyzed anonymously in a four-step process. In a first step, all responses were transcribed by a student worker. Then, they were reviewed by one researcher, as to whether they describe a situation in which a person experiences reactance. Afterward, two researchers independently clustered the responses that did describe situations where state reactance occurred according to the specific cause of state reactance in a double-blind coding procedure. As a result, both researchers ended up with six clusters of situations. The main difference between the two sets of clusters was, that researcher one included a cluster called "False Information", in which two situations were clustered, that showed incorrect information, e.g., an update that would indicate a duration 10 min for a system update, but in fact, it took more than 30 min. Researcher two created a cluster named "Bad Design", which included two situations that appeared to have followed from bad system design, such as a smart home that switches off automatic lights too early. Both of these clusters were related to the fact that they included a situation in which a smartphone would shut itself down even though the battery indicator still showed 5% charge. A third step was taken to consolidate the two sets of clusters. A small workshop with the two researchers was conducted in which the two sets of clusters were merged by discussing and resolving differences. In this step, both researchers decided to abandon the two clusters that were not present in both of the two cluster-sets, because the respective situations could be assigned to clusters that were present in both original cluster-sets. The resulting consolidated cluster-set is shown in Fig. 5.1.

5.3.2 Reduced Freedom of Choice

Seven responses could be clustered around the term *reduced freedom of choice*. This group of responses generally describe situations in which the user of a device or service is confronted with a situation in which an option of choice, one that might have been formerly available or is desirable, is not available. Examples include a Windows update that can only be postponed but not canceled completely or videos from Youtube[1] that could not be viewed because the GEMA[2] would not allow it.

5.3.3 High Effort or Costs

Four responses described situations in which the participants thought that the effort or the costs of reaching a goal were unnecessarily high. For example, one participant described a service of Technische Universität Berlin, called *Moses*. According to the description, *Moses* makes it hard to find a function that the participant was looking for because it was hidden deep in a menu which itself was hard to find in the first place. Other accounts describe an online shop that introduced fees for a certain payment method, which made the user reactant. Also, a situation with a computer update was mentioned again, this time in the context of high effort. Instead of ten minutes, the update took much longer and as it was not finished after 30 min, the participant canceled it.

5.3.4 System Errors

Five responses described situations in which system errors led to state reactance. One of those responses describes a situation in which the participant wanted to use the TV of a rented apartment, but the remote control did not work. Therefore that person

---

[1] https://www.YouTube.com.

[2] A German performance rights organization.

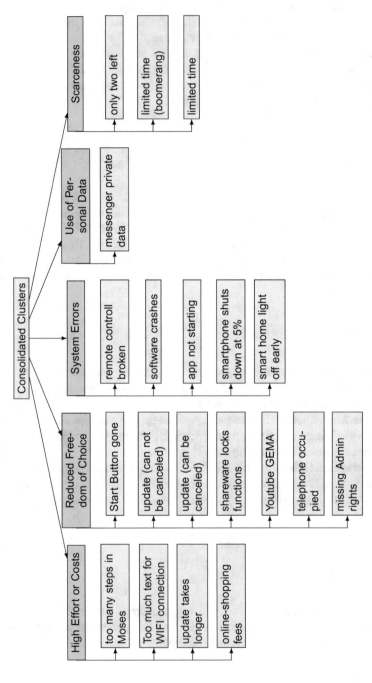

Fig. 5.1 Final set of clusters after they were consolidated by both researchers

had to get up and switch channels via the buttons on the TV itself. This account was clustered to *system errors*, but could also be counted to *high effort or costs* because the person describes that he or she had to get up in order to switch the channel. Another participant reports that a smartphone was shutting down itself even though the battery indicator showed that there was still some power left.

5.3.5 Privacy Violations and Use of Personal Data

One participant reported a situation when he or she learned that a messenger application was analyzing the content of messages and even relayed it to third parties.

5.3.6 Scarceness

Three participants reported situations in which scarceness of a product led them to buy that product. The reports describe that the scarceness was induced when the online shops advertised products as either being of limited stock or of being reduced in price for a limited time. All three participants reported that they then bought the advertised product.

5.3.7 Research Question 1.2

In order to gather more data for answering Research Question 1.2, the consequences of reactance that the participants described, for the technical device or service that was in use, were collected. In total, only five out of 20 participants reported actual consequences of state reactance for the device or service that they had used. Two of these reports stated that the participants had bought articles in an online shop as a result of state reactance. Two other participants reported that they did not use services, that triggered state reactance, anymore. Another participant reported having canceled a system update that took longer than communicated by the system. However, that person reported having continued using the system afterward.

Chapter 6
Smart Home Study—Trait Reactance

6.1 Introduction

Results from Sect. 4.2 show that evidence for trait reactance being an influential factor for interaction with technical devices and services is rather scarce. Only the study of Li and Meeds [10] showed that trait reactance could be of significance in human–computer interaction research. In order to clarify a possible influence of trait reactance on human–computer interaction further, a laboratory experiment was conducted. Theoretically, trait reactance is an attractive metric for adaptive and persuasive services, because highly reactant personalities are more likely to experience state reactance when they feel that their freedom is threatened. In this experiment, it was investigated if a person level of trait reactance can have an influence on how different interaction strategies of a technical system are perceived. Such interaction strategies can be user-adaptable and self-adaptive configuration or behavior. While self-adaptive behavior requires less action from the users and should, therefore, be more comfortable to use, it also draws control over the situation from the users toward the system itself, because the system acts proactively and adjusts itself to the situation.

Hypothesis 1 *Li and Meeds [10] showed that persons who are high in trait reactance show more add avoidance than persons who are low in trait reactance when adds are shown for the first time. Apart from this, evidence about the influence of trait reactance on human–computer interaction is scarce. It was argued in Chap. 1, that adaptive system behavior might trigger the perception of lost control over the system in their users, and that state reactance is triggered by a perceived loss of control. Also given, that people who show a high level of trait reactance can more easily enter a reactant state, then it seems likely, that people who show a high level of trait reactance are more sensitive to experience a threat to freedom or control. Hypothesis 1 therefore states that users, who score high in trait reactance will perceive a higher threat to control than users who score low in trait reactance in a self-adaptive condition.*

© Springer Nature Switzerland AG 2020
P. Ehrenbrink, *The Role of Psychological Reactance in Human–Computer Interaction*, T-Labs Series in Telecommunication Services,
https://doi.org/10.1007/978-3-030-30310-5_6

An aim of the experiment described in the following is to find out if the loss of control for the users by self-adaptive systems is also experienced as such. Also, it is investigated if such an experience is evaluated differently by personalities with high or low levels of trait reactance.

Hypothesis 2 *It is argued in Hypothesis 1 that people who show high levels of trait reactance are more sensitive to experience a threat to freedom or control. Also, they are likely to be more prone to becoming reactant and feeling the urge of protecting the threatened freedom or control. This suggests that experiencing such a threat is not desirable to highly reactant persons. Self-adaptive system behavior could pose a threat to control. Therefore, it is argued that persons who show high levels of trait reactance are experiencing less pleasure when interacting with the self-adaptive system.*

Hypothesis 3 *Hypotheses 1 and 2 address measures of consequences of adaptive system behavior on the experience of the subjects, but this does not include direct consequences for the interaction behavior. One indicator of state reactance is source derogation [4, 12]. In a human–computer interaction experiment, source derogation could be accomplished by a more negative rating of the technical system that poses a freedom threat. Hypothesis 3 therefore states that the self-adaptive system will pose a higher threat to freedom and that this will result in a lower rating of acceptability[1] of the self-adaptive system among the group of participants who show high levels of trait reactance.*

6.2 Methods

In order to test if trait reactance is a factor that has any influence in human–computer interaction, it was implemented as a metric in an experiment that investigated differences between self-adaptive and user-adaptable system behavior in a smart home setting.

Measurement of trait reactance was accomplished with the HPRS [5–7] with a unidimensional configuration that was suggested by Yost [15] on a five-point Likert-scale [11]. The perceived freedom threat was operationalized with the factor dominance of the Self Assessment Manikin (SAM) [1]. This questionnaire also contains the factor Pleasure, which was used to assess the pleasure that the participants feel.

Acceptability was operationalized with the C1 dimension, that Möller et al. identified in a questionnaire according to ITU-T recommendation P.851 [9, 13]. The questionnaire will be called P.851 in the following.

---

[1]In theory, acceptance of a device or service that triggers state reactance would be expected to decrease. Acceptance could be measured in long-term studies, by observing how much a product is purchased and used. However, in a laboratory experiment, acceptance can not be measured directly. Therefore, acceptability shall be used here. Acceptability can be regarded as the potential of a product to gain acceptance.

The interactive systems were solely designed and build for this experiment and manually controlled by a person in a Wizard-of-Oz scenario. The systems were realized in an autonomous (self-adaptable) and in a controllable (user-adapted) version. Both versions had an equal functional range and each participant interacted only with either the self-adaptable or the user-adapted version. The independent variable adaptivity was therefore a between-participants variable. The self-adaptive version changed its behavior according to the context (e.g., the subject's gender and current task), while the behavior of the user-adaptable version was completely controlled by the subject. The system was a smart home environment that consisted of several parts. The first part was a simple spoken dialogue system. It started off with a robotic-like voice that was created using Mary TTS and with the voice *bits3-hsmm*, which was altered with the parameters *effect_Robot_parameters = amount:100.0* and *effect_Robot_selected = on*, so that it would sound machine-like. The spoken dialogue system was also able to speak with a male voice, using the unaltered voice *bits3-hsmm* and with a female voice, using the unaltered voice *bits1-hsmm* [3]. The system voice could adapt or be adapted to the gender of the user. If the user was male, the system would use a male voice, if the user was female, the system would use a female voice. The second part was a music system that could lower the volume if somebody would speak. In the self-adaptive version, this was happening automatically. In the user-adaptable version, it had to be triggered by a voice command. The third part was a smart reading light. The self-adaptive version detected if the room was bright enough to read and switched itself on if this was not the case anymore. The user-adaptable version had to be switched on by a voice command.

All automatic speech recognizer functions were substituted by a human which remotely operated all devices from a separate room in a Wizard-of-Oz setting. In such a setting, a real person (the wizard) listens to the subject's utterances (via a remote microphone) and triggers the related responses. In this case: playing prompts, toggling light and controlling the music player. The wizard simulated a perfect speech recognizer that always understands the subject's utterances and the communicated intention. This was done to ensure that only conceptual properties of the system were taken into account by the subjects, instead of technical shortcomings. The participants were not informed about the wizard prior to the experiment and thought to interact with a fully functional system. A visualization of the experimental setup can be seen in Fig. 6.1.

6.2.1 Participants

44 persons took part in the experiment. Most of them were recruited by flyers and posters in supermarkets and on traffic lights, as well as with a test person database within TU-Berlin. All participants that were not employed by TU-Berlin received ten Euro as reimbursement for taking part in the experiment. During data analysis, the dataset of one participant was dropped because it represented an extreme outlier.

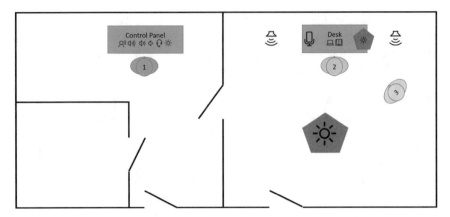

Fig. 6.1 Experimental setup and location. Person 1: Wizard, Person 2: Participant, Person 3: Experimenter

6.2.2 Procedure

After arrival, all participants were greeted by the experimenter and let to the room in which the experiment would take place. The wizard was sitting in another room that was connected to the experiment room via a door that was only open by a little gap. The participants could not see the wizard.

Each participant got an introduction sheet and a form of consent that had to be signed. Also, the participants were informed by the experimenter that it would not matter if the participants rated the system that they interacted with good or bad, but that only the honest rating of the participants was of interest. This was done to reduce a social desirability effect that could have had resulted in less valid results. Afterward, they were asked to fill out the refined version of the HPRS [5] and the SAM scale[2] [1].

Then, the main part of the experiment followed. Prior to each task, the examiner read a scenario description with a background story to the subject. It introduced the participant to an imaginary situation at home and specified implicitly the task to be carried out. Afterward, the participant carried out the task by interacting with one of the two versions of the system. The scenario description served the purpose of distracting the participant from the fact that he or she was participating in a laboratory experiment. Since the systems that were used in the experiment were part of a simulated smart home, aiding the subject's feeling of being home was thought to increase the validity of the collected data.

In the first part, the participants interacted with a spoken dialogue system, which was part of the smart home. Both versions could talk to the participants with either a robot-like, a female or a male voice. The speech output was realized with predefined prompts, uttered with MARY Text-to-Speech [3]. Interaction with the spoken

---

[2]This was done to get a baseline (pre) measurement for dominance and pleasure.

dialogue system started with the system introducing itself in a robot-like voice. Afterward, the user-adaptable system would ask the participant if it should use a male or a female voice. The participant then answered vocally which voice the system should use. Then, the system gave a confirmation of the new voice being used in that particular voice. The self-adaptive system asked the participants if they were male or female. After the participants responded, the system proactively changed its voice to a voice matching the subject's gender (male-male, female-female)[3] and communicated the change in a voice message. Following this, the participants interacted with the other functions of the mock-up smart home system. Those contained a music player and a reading lamp that could be controlled via voice commands.

The scenario description told the participants that they want to hear music after the system's activation. The participants then started the music with a voice command and were instructed to inform the examiner before switching off the music player whenever they wanted. The self-adaptive version started the music with very low volume and increased the volume automatically to room volume level after one second. This was the first hint to emphasize that the music volume was automatically controlled. Each time the participants or examiner talked, the volume was turned down automatically until talking stopped. In the user-adaptable condition, the music also started very quietly, but there was no automatic raising to room volume level nor decrease while attendees talked. Only participants of the user-adaptable condition were informed about the possibility to control the music volume by voice. Some participants of the self-adaptive condition tried it anyway, without being explicitly informed about that possibility. The wizard followed the command in such cases.

In the task with the reading lamp, participants were instructed to read an article in a print magazine. The background story stated that they were living in a shared flat and that the examiner would act as their flatmate. In the beginning of the interaction, the room was lid by a ceiling lamp, but the examiner switched off the ceiling lamp two minutes after the participants had started reading the article. The experimenter then switched it on again, after two additional minutes. Participants in the user-adaptable condition were informed that they can control the reading light by speech, and all of them did, once the ceiling lamp was switched off. Participants in the self-adaptive condition were not informed how to control the reading light, but the system toggled it automatically according to the status of the ceiling lamp. If the ceiling lamp was off, the reading light turned on automatically until the ceiling lamp was switched on again. After the interaction with the system, all participants were asked to answer several questionnaires, including the P.851 [9] and the SAM [1].

---

[3] A gender-neutral or gender-unspecific option was not included and not asked for during the course of the experiment.

6.3   Results

Visual inspection of the data using IBM SPSS 25 [8] revealed an outlier for the factor Pleasure. The dataset of this participant was removed for analysis because it represented an extreme outlier according to the Tukey Fences [14] criterion. A more detailed description of the Tukey Fences criterion is provided in Sect. 10.4.

The measurement of Hong's Psychological Reactance Scale was calculated according to [2, 15], using items 1, 6, 9, 10, 11 and 12 as a single factor for trait reactance. The trait reactance score was then used to divide all participants into two groups, according to whether their trait reactance score was above or below the median (3.055). The resulting participant groups over the two conditions can be viewed in Table 6.1. The *low trait reactance* group had an average trait reactance score of 2.510 (SD = 0.299) and the *high trait reactance* group had an average trait reactance score of 3.582 (SD = 0.374). Descriptive statistics for dominance (threat to control), pleasure, and acceptability for the two groups and the two conditions are shown in Tables 6.2, 6.3 and 6.4 and Figs. 6.2, 6.3 and 6.4.

Table 6.1 Number of participants with high and low trait reactance scores in the two conditions

	High trait reactance	Low trait reactance
User-adaptable	N = 11	N = 11
Self-adaptive	N = 10	N = 10

Table 6.2 Dominance as measured by SAM after interacting with the smart home

	High trait reactance		Low trait reactance	
	Mean	Std.	Mean	Std.
User-adaptable	6.64	1.362	5.0	1.342
Self-adaptive	4.70	1.418	5.55	1.753

Table 6.3 Pleasure as measured by the SAM after interacting with the smart home

	High trait reactance		Low trait reactance	
	Mean	Std.	Mean	Std.
User-adaptable	7.272	1.272	7.0	1.183
Self-adaptive	5.900	1.287	7.0	1.483

Table 6.4 Acceptability as measured by the P.851 after interacting with the smart home

	High trait reactance		Low trait reactance	
	Mean	Std.	Mean	Std.
User-adaptable	2.671	0.313	2.375	0.407
Self-adaptive	2.625	0.433	2.671	0.327

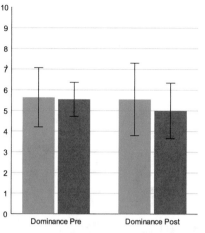 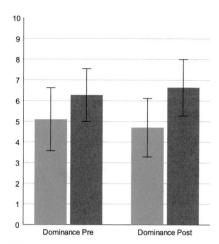

(a) Dominance ratings of lowly reactant participants in the two conditions.

(b) Dominance ratings of highly reactant participants in the two conditions.

Fig. 6.2 Mean ratings and standard deviations of dominance in the two conditions self-adaptive (blue) and user-adaptable (red) between the two groups of lowly (left) and highly (right) reactant subjects

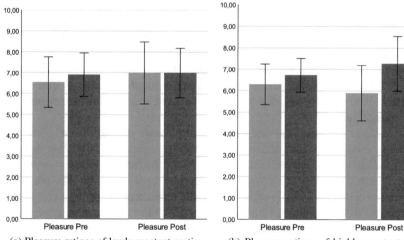

(a) Pleasure ratings of lowly reactant participants in the two conditions.

(b) Pleasure ratings of highly reactant participants in the two conditions.

Fig. 6.3 Mean ratings and standard deviations of pleasure in the two conditions self-adaptive (blue) and user-adaptable (red) between the two groups of lowly (left) and highly (right) reactant subjects

Fig. 6.4 Mean ratings and standard deviations of acceptability in the two conditions self-adaptive (blue) and user-adaptable (red) between the two groups of lowly and highly reactant subjects

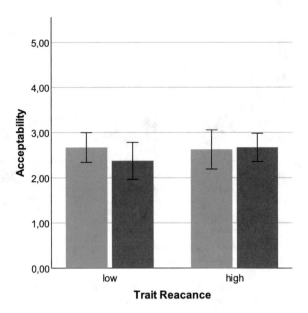

Independent samples t-tests were used to test if the observed differences between the highly reactant and the lowly reactant personality groups in the baseline measurement and the post-interaction measurements of Pleasure and Dominance are significant for the self-adaptive and the user-adaptable condition. Levene-Tests for homogeneity of variance were performed prior to all independent samples t-tests described in the following. They resulted in non-significant results in every case, suggesting that homoscedasticity can be assumed and an independent samples t-test can be performed.

The independent samples t-test showed that pleasure was significantly lower in self-adaptive condition than in the user-adaptable condition with $t(19) = 2.456$, $p < 0.05$ for highly reactant participants. Ratings for the factor dominance were significantly smaller in the self-adaptive condition, compared to the user-adaptable condition with $t(19) = 3.191$, $p < 0.01$ for highly reactant participants, meaning that those people felt less dominant when interacting with the self-adaptive system.

No significant differences of acceptability were observed between the two conditions in the group of highly reactant participants. Also, no significant differences of pleasure, dominance, or acceptability were observed between the two conditions in the group of lowly reactant participants.

References

1. Bradley, M.M., Lang, P.J.: Measuring emotion: the self-assessment manikin and the semantic differential. J. Behav. Ther. Exp. Psychiatry **25**(1), 49–59 (1994)
2. Brown, A.R., Finney, S.J., France, M.K.: Using the bifactor model to assess the dimensionality of the hong psychological reactance scale. Educ. Psychol. Meas. **71**(1), 170–185 (2011)
3. DFKI GmbH: The Mary text-to-speech system (marytts) (2015). Accessed 10 Feb 2016. http://mary.dfki.de/
4. Grandpre, J., Alvaro, E.M., Burgoon, M., Miller, C.H., Hall, J.R.: Adolescent reactance and anti-smoking campaigns: a theoretical approach. Health Commun. **3**, 349–366 (2003)
5. Hong, S.M., Faedda, S.: Refinement of the Hong psychological reactance scale. Educ. Psychol. Meas. **56**(1), 173–182 (1996). https://doi.org/10.1177/0013164496056001014. URL https://doi.org/10.1177/0013164496056001014
6. Hong, S.M., Page, S.: A psychological reactance scale: development, factor structure and reliability. Psychol. Rep. **64**(Suppl 3), 1323–1326 (1989). https://doi.org/10.2466/pr0.1989.64.3c.1323. URL https://doi.org/10.2466/pr0.1989.64.3c.1323
7. Hong, S.M.: Hong's psychological reactance scale: a further factor analytic validation. Psychol. Rep. **70**(2), 512–514 (1992). https://doi.org/10.2466/pr0.1992.70.2.512. URL https://doi.org/10.2466/pr0.1992.70.2.512
8. IBM Corp.: IBM SPSS statistics for windows. IBM Corp., Armonk, NY (2015)
9. International Telecommunication Union: subjective quality evaluation of telephone services based on spoken dialogue systems. Technical Report, Supplement 851 to P-Series Recommendations, International Telecommunication Union, Geneva, Switzerland (2003)
10. Li, C., Meeds, R.: Factors affecting information processing of internet advertisements: a test on exposure conditions, psycholgical reactance, and advertising frequency. In: Proceedings Conference on the American Academy of Advertising, pp. 93–101. American Academy of Advertising, Austin (2007)
11. Likert, R.: A technique for the measurement of attitudes. Arch. Psychol. (1932)
12. Merz, J.: Fragebogen zur messung der psychologischen reaktanz. Diagnostica **29**(1), 75–82 (1983)
13. Möller, S., Smeele, P., Boland, H., Krebber, J.: Evaluating spoken dialogue systems according to de-facto standards: a case study. Comput. Speech Lang. **21**, 26–53 (2007)
14. Tukey, J.W., Cromwell, L.: Exploratory data analysis, Addison-Wesley Publishing Company, **1** (1977) ISBN-13: 978-0-201-07616-5
15. Yost, A.B., Behrend, T.S., Howardson, G., Badger Darrow, J., Jensen, J.M.: Reactance to electronic surveillance: a test of antecedents and outcomes. J. Bus. Psychol. (2018). https://doi.org/10.1007/s10869-018-9532-2. URL https://doi.org/10.1007/s10869-018-9532-2

Chapter 7
Intermediate Discussion on the Relevance of Psychological Reactance for Human–Computer Interaction

7.1 Introduction

In order to shed light on the relevance of psychological reactance for human–computer interaction, a literature search a qualitative survey and a laboratory experiment were conducted. The implications and shortcomings of these studies are discussed in the following.

7.2 Research Question 1

Research Question 1 asked whether psychological reactance is of relevance for the domain of human–computer interaction, at all. In order to answer this question, a literature search was performed. In total, this literature search found twelve papers in which psychological reactance was investigated in the context of human–computer interaction either as a state or as a trait. The papers were then analyzed regarding the way that psychological reactance was investigated. Specifically, if they had produced evidence that either state reactance was triggered by a technical device or service, or that trait reactance influenced interaction with technical devices or services. For state reactance, the results point in the direction that state reactance can be triggered by a technical device or service. Out of the twelve papers, two produced direct evidence by actually measuring state reactance, and seven produced indirect evidence by observing behavioral patterns that indicate state reactance, such as source derogation or boomerang effects.

For trait reactance, the results were less convincing. Only one paper was able to present direct evidence for trait reactance having an influence on human–computer interaction. Also, the observed differences in add avoidance behavior on websites for high and low trait reactance groups show inverted effects after repeated presentations. The authors Li and Meeds come up with the explanation that repeated add avoidance

© Springer Nature Switzerland AG 2020
P. Ehrenbrink, *The Role of Psychological Reactance in Human–Computer Interaction*, T-Labs Series in Telecommunication Services,
https://doi.org/10.1007/978-3-030-30310-5_7

takes many resources and that persons with high trait reactance consider this an intrusion (and therefore stop add avoidance after some time) [7]. Still, the inversion of the effect between participants with high and low trait reactance cast some doubt on the assumption that it is best explained by trait reactance.

Because of the scarce evidence for trait reactance playing a role in human–computer interaction, a laboratory experiment was conducted to test, if differences in human–computer interaction exist between participants with high and low trait reactance. As it was argued in Chap. 1, adaptive system behavior transfers some control from users to the system itself and human-likeness of the system might even increase this effect. Therefore, system behavior was introduced as an independent variable with the conditions self-adaptive system behavior and user-adaptable system behavior of a smart home system. In order to include a human-like component to the smart home system, interaction was realized via speech interaction and the voice that the smart home system used after it was configured by the participant was a human-like one.

Hypothesis 1 claimed, that participants with high trait reactance might perceive a higher threat to freedom by the self-adaptive smart home system than participants with low trait reactance. The threat to freedom was operationalized with the dominance dimension of the SAM. This dimension measures how dominant or autonomous a participant feels in a given situation. If a participant receives a threat to freedom or control, it should lower the perceived dominance of the subject. Perceived dominance can, therefore, be regarded as an indirect measure of threat to freedom or control. The observed data shows that highly reactant personalities feel significantly less dominant in the self-adaptive condition, than in the user-adaptable condition, which confirms Hypothesis 1. The difference is quite severe, as can be seen in Fig. 6.2b. There is also a difference in the dominance ratings between the two conditions in the baseline measurement. Although this difference is not significant, it casts some doubt on the validity of the post-interaction measurement and suggests, that there might have been an imbalance between the participant groups of the two conditions. On the other hand, no such difference could be observed in the baseline measurement of the lowly reactant subjects, as can be seen in Fig. 6.2a.

Hypothesis 2 claimed, that participants with high trait reactance might perceive less Pleasure in the self-adaptive condition than in the user-adaptable condition. Also, this hypothesis can be confirmed. The observed data shows that highly reactant participants felt significantly less Pleasure in the self-adaptive condition, compared to the user-adaptable condition. For Pleasure, no considerable difference could be observed in the baseline measurement of highly reactant subjects. Also, lowly reactant participants did not show significant differences in their ratings of Pleasure, which points to differences in the experience of threats to freedom or control to the two trait reactance groups. The results for pleasure are illustrated in Fig. 6.3.

Evidence indicates that participants who show high levels of trait reactance experience self-adaptive system behavior significantly different than user-adaptable system behavior, indicating that self-adaptive system behavior really poses a threat to experienced freedom of control. But the experienced freedom threat does not necessarily mean that there are negative consequences for the technical system that is in use.

Table 7.1 Overview of the number of papers concerned with a group of situations in which reactance effects occur and the respective number of responses from the survey

Situation	Literature	Survey
Persuasive attempts	5	0
Use of personal data	2	1
Reduced freedom of choice	1	7
Forced response	1	0
Obstructions	1	0
High effort or costs	1	4
Scarceness	1	3
System errors	0	5

Hypothesis 3 predicted that highly reactant participants would rate the self-adaptive smart home lower in terms of acceptability. However, no significant difference was observed. As Fig. 6.4 shows, ratings by the highly reactant participants are similar for both systems. Lowly reactant participants rated the user-adaptable system a little higher on average, but this difference is not significant. Therefore, Hypothesis 3 cannot be confirmed.

7.3 Research Question 1.1

In order to answer Research Question 1.1, which asks in what situations psychological reactance relevant would be of relevance for human–computer interaction, a literature search, and a qualitative survey were conducted.

The literature search produced a total of twelve papers that covered seven types of situations in which reactance played a role, either as a state or as a personality trait. Overall, it became apparent, that literature covered the topic of reactance in the context of human–computer interaction quite widely. Only persuasive attempts and use of personal data received the attention of multiple papers.

Situations Described in Literature and the Survey

Comparing the results of the literature search and the survey reveals a discrepancy in representation between the observed distribution of different situations in which reactance occurs and the concentration of research. The reported situations and the number of occurrences are shown in Table 7.1.

Persuasive Attempts

State reactance as a result of persuasive attempts was covered the most in the reviewed literature. This might indicate that psychological reactance is especially relevant for the domain of persuasive technology. Persuasive technology is technology that aims at changing the behavior or attitudes of users [11]. Changing behavior of a person necessarily implies taking control over that person to some extent. Hence, the person will suffer a loss of control and probably experience the resulting threat to behavioral freedom, which is the mechanism that triggers state reactance. Torning and Oinas-Kukkonen describe persuasive technology the following way [11]:

> In using technology as a vehicle of persuasion, we touch upon a central part of being human, namely intentional communication. Whenever we communicate deliberately with a clear purpose and outcome in mind, we are engaging in persuasion.

This description indicates, that persuasive technology is technology that communicates to humans in order to reach a specific goal, namely behavior change. This is highly related to the definition of intelligent systems that was introduced in Chap. 1, defining intelligent systems as systems of which their functionality suggest that they are agenda driven. It is argued in Chap. 1, that intelligent systems might be more prone to trigger state reactance than systems which are not regarded as intelligent. In this light, it seems very likely that one class of situations that trigger state reactance is when technical systems commit persuasive attempts. On the other hand, persuasive attempts have not been described in the expert survey. This could be due to relatively low or even no exposure of the experts to persuasive systems. Even though (adaptive) online advertisement in different forms is encountered regularly on websites, recommender systems might not consciously be regarded as persuasive attempts or simply do not cause sufficient threats to freedom to trigger state reactance.

Use of Personal Data

Use of personal data was a trigger for state reactance in the survey report of one subject. Also, two studies were concerned with the use of personal data in the context of human–computer interaction. In both studies, participants indicated a lower intention to use the version of the service that used personalized information. Apart from the human–computer interaction domain, use of personal data has been described as a trigger for state reactance, before. Already Merz reported that highly reactant personalities complained about the use of personal data more often [8, p.80]. Also, Yost et al. conducted a study in which they investigated the relationship of trait reactance, state reactance, and workplace surveillance [12]. They found a positive correlation between trait reactance and the perceived invasion of privacy by surveillance at workplaces. They also found that state reactance is positively related to the extent to which surveillance is perceived as an invasion of privacy [12]. It seems therefore very likely that use of personal data in human–computer interaction can cause psychological reactance.

Reduced Freedom of Choice

Reduced freedom of choice is the main mechanism that triggers state reactance [1, 2]. Also, one of the identified papers dealt with reduced freedom of choice in the human–computer interaction domain [9] and found that this mechanism also holds for interface preferences of software. Despite the relatively small coverage of this state reactance-trigger in human–computer interaction literature, a direct reduction of the freedom of choice seems to be encountered frequently among users. Out of the twenty valid responses that were analyzed for the qualitative survey, 7 reported of situations in which their freedom of choice was reduced by the technical system they were using.

Forced Response and Obstructions

Forced responses and obstructions by technical systems were investigated by a single paper, each. Stieger et al. claimed that forced responses to online questionnaires were causes of state reactance, while Edwards et al. investigated pop-up adds and figured that these were regarded as obstructions during web surfing. However, neither forced responses nor obstructions were named during the survey, which suggests that they play minor roles as triggers for state reactance in human–computer interaction.

High Effort or Costs

One of the identified papers investigated trait reactance in connection with online add avoidance. Li and Meeds concluded that highly trait reactant participants drop their strategy of add avoidance after some repetitions because of the higher cost of mental ressources [7]. Additionally, four participants from the survey mentioned high effort or costs as a trigger that caused state reactance when they were interacting with technical systems. This seems valid, regarding that e.g., higher costs of time can be perceived as equivalent to a higher threat to freedom.

Scarceness

Scarceness was investigated in one paper from the domain of human–computer interaction. Zheng et al. could show that users of online shops are sensitive to scarceness advertisement [13]. Also, three participants from the survey reported that they had been influenced by scarceness marketing. The number of responses concerning scarceness in the survey is probably biased by the fact that scarceness was part of the introductory material on psychological reactance that the participants received. Scarceness might be a trigger for state reactance, but its relevance for human–computer interaction seems to be restricted to marketing. Therefore, it is regarded as being of no further relevance for this work.

System Errors

None of the identified papers studied the effects of system errors on state reactance. In contrast, five participants of the survey reported having turned reactant after a technical system produced an error of some sort. It could be argued that developers of technical systems try to reduce the number of errors that the systems produce, anyway. Nevertheless, understanding the factors that play a role in such situations might enable designers and developers to reduce the negative effects that are following errors that can not be or are not avoided.

7.4   Research Question 1.2

Research Question 1.2 asked about the consequences that state reactance has for the interaction with the system that triggers state reactance. To answer this question, the identified literature was analyzed in terms of identified consequences. Also, participants in the survey were asked to describe the consequences that the experience of state reactance had for interaction with the technical system that triggered it. The results from the literature are shown in Table 4.2, the results from the survey are described in Sect. 5.3.7. Three of the collected studies described boomerang effects. A boomerang effect is an effort of doing the opposite of what is requested, or presumably requested by the system. Also, three other studies reported avoidance behavior or that the participants ignored the services that triggered state reactance. Further studies indicated that the participants who experienced state reactance had stopped interacting with the system or showed less intention to use the system in the future. One study reported less acceptance for the interface that was forced upon the subjects. Responses from the expert survey are rare. Only five out of the twenty responses that were analyzed reported any consequences for the systems that were the cause of the state reactance experience. Two of the described consequences were that the participants had bought products as a result of scarceness advertisement. Two other responses were rather drastic. Both participants reported that they did not use the technical system anymore, because of the incident that triggered state reactance. One such situation was the use of personal data, the other one was reduced freedom of choice. The last reported consequence was that a user aborted the update that caused state reactance, but that he or she had continued it at a later point.

7.4.1   Smart Home Study: Acceptability

The smart home study included a measure of acceptability from the P.851 questionnaire [6]. Despite the measurements of Dominance, which indicate that the self-adaptive system posed a higher threat to freedom or control to the subjects, and the fact that this is supported by the measurements of Pleasure, no difference in acceptability was observed. For state reactance, this could either mean that it was triggered, but had no effect on acceptability, or, it was not triggered. If it was not triggered, the freedom threat might not have been existent, or it was not high enough.

Figure 6.2b shows that there was already a difference in the dominance rating of the two system conditions prior to interaction. Since the participants were assigned at random to the system groups, this difference is probably by chance, also it is not significant. The observed difference in Dominance after interaction with the self-adaptive system was significant at a $p < 0.01$ level, meaning that the chances of it being by chance are less than 1%. Chances are that Dominance is not an adequate operationalization for freedom threat. If state reactance was triggered by the self-adaptive system remains inconclusive after the experiment.

7.5 Methodological Criticism

The question of the qualitative survey was formulated as a forced response in that it did not ask if participants had ever experienced state reactance, but how they had experienced state reactance while interacting with technical devices or services. It could be argued that the formulation of the question could have resulted in invented responses. This is a valid refutation. But firstly, the answers would still provide a hint for what situations are likely candidates to trigger state reactance. And secondly, the answers were not used to answer the question if state reactance can be triggered by technical devices or services.

7.5.1 Completeness

Neither results of the literature search, nor of the qualitative survey can claim completeness. However, it is unlikely that the literature search has missed a substantial amount of literature, since it included large international search engines, such as Google Scholar and also cross-references of the identified studies did not reveal other work. The qualitative survey, on the other hand, shows some methodological shortcomings. First, the participants had to get an introduction to psychological reactance, which likely resulted in biased results. Also, only 21 participants participated in the study. An interesting outcome is, that the described situations can be assigned to a relatively limited number of categories, such as system errors and high effort or cost. Both categories were not mentioned in the introduction to psychological reactance. Apart from the use of personal data, all categories include at least three reports, suggesting that the collected categories might indeed be the ones that are the most representative within the population of situations.

7.5.2 Trait Reactance

Trait reactance was measured by a set of items from Hongs's psychological reactance Scale [5]. It was already described in detail in Sect. 2.3.1.4, that there has been a lot of debate about how to use the questionnaire [3, 4, 10]. It was decided to use the set of items that were proposed by Yost et al. who decided on this set on the basis of previous work, that compared the competing measurement models [3, 12].

References

1. Brehm, J.W.: A Theory of Psychological Reactance. Academic Press, New York (1966)
2. Brehm, S.S., Brehm, J.W.: Psychological Reactance: A Theory of Freedom and Control. Academic Press, New York (1981)
3. Brown, A.R., Finney, S.J., France, M.K.: Using the bifactor model to assess the dimensionality of the hong psychological reactance scale. Educ. Psychol. Meas. **71**(1), 170–185 (2011)
4. Hong, S.M., Faedda, S.: Refinement of the Hong psychological reactance scale. Edu. Psychol. Meas. **56**(1), 173–182 (1996). https://doi.org/10.1177/0013164496056001014
5. Hong, S.M., Page, S.: A psychological reactance scale: Development, factor structure and reliability. Psychol. Rep. **64**(Suppl 3), 1323–1326 (1989). https://doi.org/10.2466/pr0.1989.64.3c.1323
6. International Telecommunication Union: Subjective quality evaluation of telephone services based on spoken dialogue systems. Technical report, Supplement 851 to P-Series Recommendations, International Telecommunication Union, Geneva, Switzerland (2003)
7. Li, C., Meeds, R.: Factors affecting information processing of internet advertisements: a test on exposure conditions, psycholgical reactance, and advertising frequency. American Academy of Advertising. Conference, Proceedings, pp. 93–101. American Academy of Advertising, Austin (2007)
8. Merz, J.: Fragebogen zur messung der psychologischen reaktanz. Diagnostica **29**(1), 75–82 (1983)
9. Murray, K.B., Häubl, G.: Freedom of choice, ease of use, and the formation of interface preferences. MIS Q. **35**(4), 955–976 (2011)
10. Shen, L., Dillard, J.P.: Psychometric properties of the hong psychological reactance scale. J. Pers. Assess. **85**(1), 74–81 (2010)
11. Torning, K., Oinas-Kukkonen, H.: Persuasive system design: state of the art and future directions. In: Proceedings of the 4th International Conference on Persuasive Technology, Persuasive 2009, pp. 30:1–30:8. ACM, New York, NY, USA (2009). https://doi.org/10.1145/1541948.1541989
12. Yost, A.B., Behrend, T.S., Howardson, G., Badger Darrow, J., Jensen, J.M.: Reactance to electronic surveillance: a test of antecedents and outcomes. J. Bus. Psychol. (2018). https://doi.org/10.1007/s10869-018-9532-2
13. Zheng, X., Liu, N., Zhao, L.: A study of the effectiveness of online scarce promotion-based on the comparison of planned buying and unplanned buying. In: Proceedings of the 12th Wuhan International Conference on E-Business, pp. 247–257 (2013)

Part III
Measurement of State Reactance

Chapter 8
Reactance Scale for Human–Computer Interaction

8.1 Introduction

In order to conduct meaningful research on psychological reactance, a metric was needed that is able to quickly and reliably assess a person's level of state reactance. Also, such a metric should be easily integrable into an existing test batterie for user experience and usability evaluation experiments. A commonly used method of assessing state reactance is the mixed-method approach from Dillard and Shen [10], which combines a short questionnaire and a thought-listing task to assess a person's level of state reactance. While this method is a valid approach, it is not very usable in itself, because it requires a trained experimenter to perform the thought-listing task. The thought-listing task not only requires the experimenter to interact with the subjects, thereby introducing the risk of certain biases like gender bias or social desirability, it also constitutes a change of modality.[1] Therefore, a state reactance metric in the form of a questionnaire was desirable for further research. There are currently other questionnaires that are able to assess state reactance [26, 29, 31]. Those questionnaires are likely not useful in the field of human–computer interaction research for reasons explained in Sect. 2.2.4. The development of the Reactance Scale for Human–Computer Interaction (RSHCI), that is described in the following sections has been published in [12]. The author of this thesis has planned, conducted and analyzed the described research. The texts that are used in this thesis to describe the development are derived from this publication.

---

[1] Anger is assessed via a Likert-Type questionnaire, while Though Listing is a qualitative task that is performed in interaction with another person.

© Springer Nature Switzerland AG 2020
P. Ehrenbrink, *The Role of Psychological Reactance in Human–Computer Interaction*, T-Labs Series in Telecommunication Services,
https://doi.org/10.1007/978-3-030-30310-5_8

8.2   Research Question: How Can State Reactance Be Measured?

After the concept of state reactance has been introduced in Sect. 2.2, the prototypical situation in which state reactance might be triggered were identified along with some of the potential consequences. In order to investigate state reactance and its consequences for human–computer interaction further, an adequate measurement technique was needed. As it was already concluded in Sect. 2.5, there was no adequate measurement technique to date, that can easily be applied in laboratory experiments and online studies. This is because the existing techniques either required examiner intervention or extensive reformulation of the items to fit the current situation. Both properties are obstacles to the evaluation of technical systems in the laboratory or online, because, examiner intervention holds the danger of social desirability effects and reformulation involves always the danger of questionable validity.

Research Question 2   *In order to collect valid data about state reactance in laboratory experiments, as well as in online studies, a self-report based measurement tool is needed that can measure state reactance. Research Question 2 therefore asks: How can state reactance be assessed in laboratory experiments and online studies without the need of examiner intervention and extensive reformulation.*

Ideally, a measurement technique should have the form of a self-report questionnaire that does not need many reformulations. This has the advantage that it can easily be introduced to the standard set of questionnaires that are used during usability evaluation and also does not need the intervention of an examiner during online studies. A generalizable questionnaire would not only be useful in the current line of work but would also enable practitioners to easily check their implementations for whether they might induce state reactance in their users.

8.3   Item Generation

The development of the RSHCI started with generating a pool of items that was to serve as a starting point for further refinement of the scale by statistical methods. Before actual items could be generated, a set of phrases that could be used for item generation was collected from three different sources.

Existing Questionnaires

Existing questionnaires that were used for assessment of psychological reactance were the first source of phrases that could be used to access state reactance. Those questionnaires were the anger scale [8–10], the FBMPR [27] and the HPRS [16, 17]. The HPRS was developed, based on a translation of the questionnaire FBMPR [16, 27, 32],[2] but still, both questionnaires were used in this step. This was done because

---

[2]For a detailed overview refer to Sect. 2.3.1.4.

Hong's questionnaire is not only a subset of items from Merz's questionnaire but also contains altered versions of some of the items. One example for this is item seven of Merz's scale: "Frei u. selbstständige Entscheidungen zu treffen, ist mir wichtiger als den meisten anderen Menschen" [27]. This item could be translated as "Making free and independent decisions is more important to me than it is for most other people" [32], but the corresponding item in Hong's scale is "I become frustrated when I am unable to make free and independent decisions" [16]. Hong added an emotional aspect to this item that could have been important for the further development of the RSHCI questionnaire. Two researchers extracted phrases that could be formulated to address a specific trigger (of state reactance) from the above-mentioned questionnaires. The procedure resulted in a collection of 15 different phrases.

Anonymous User Comments on Websites

Another source of phrases and formulations that might hint at state reactance were anonymous user comments on websites. Two researchers collected formulations and phrases from the user comments, mainly of the websites http://www.zeit.de, which is a German news website and http://www.heise.de, which is a German news website specialized at technology. Anonymous user comments were chosen because they are less likely to be influenced by social desirability or social anxiety effects, which could influence the formulation of phrases. Joinson found that people are biased less by social anxiety and social desirability if they are communicating anonymously via the internet [23]. Therefore, anonymous user comments might provide less filtered insight into people's thoughts. Phrases from the comments of the websites were selected if they expressed either Anger, negative cognitions or a perceived threat to freedom of the users. anger and negative cognitions are the two components of state reactance, as proposed in the intertwined model by Dillard and Shen [10]. The phrases were collected independently by the two researchers. Afterward, all phrases were discussed and similar formulations consolidated. Collecting phrases from anonymous user comments resulted in 15 additional phrases.

Experts

The last source of items was a brainstorming session of six researchers who were all experts in the field of human–computer interaction and knew the concept of psychological reactance. The brainstorming session produced 22 unique phrases.

In the next step, all phrases from the three sources were reviewed by two researchers. They removed double phrases or phrases that could be offensive to some people. In total, 52 phrases were produced by the three sources, 32 of which were kept for item generation after the review process. Afterward, the two researchers used the 32 phrases to formulate a pool of 37 items. The items were formulated in a way that they address a specific object that could be the trigger of reactance or the reason for a perceived threat to freedom. Following suggestions from the brainstorming session, the items included positive and negative formulations of some of the collected phrases. The items can be viewed in Table.

8.4   Item Selection—Smart Home Online Study

Item selection was performed in a dual-method process. Data from an online experiment was used to select an initial set of items for each factor of the questionnaire via maximum likelihood factor analysis. The resulting factors were then refined using structural equation modeling.

8.4.1   Participants

All participants were recruited via the Prometei participant database of Technische Universität Berlin. At the time of the study, about 3000 persons were registered in that database. To get as many datasets as possible, the maximum number of participants to was set to 3000 persons and no restrictions regarding gender, occupation or age were set. Three times 50 Euro were drawn under all participants as an incentive to take part in the experiment. 448 persons started the experiment and 395 of them completed the whole procedure and all candidate items for the questionnaire. The only demographic information about the participants that was assessed was age. The mean age was 29.32 years (SD = 9.66 years). In order to participate in the drawing and to inhibit repeated participation in the experiment, users had to provide their email address and name after the study was finished.

8.4.2   Stimuli

In order to create variance in the data, the study contained two conditions. One condition was designed to induce state reactance (high-threat) and one condition was designed to avoid state reactance (low-threat). This was done by applying techniques from literature, including forceful language [13] to emphasize reduced freedom [4, 5] for the high-threat stimulus and the *but you are free to accept or to refuse*-technique [14] for the low-threat stimulus.

Forceful Language

Forceful language has often been used in psychological reactance research to induce a freedom threat and thereby state reactance. Forceful language uses a direct, commanding tone, such as "you absolutely must do this", to persuade a person into doing something. Its effectiveness in indicating state reactance has been shown in several studies [13, 28, 30].

The "But You are Free to Accept or to Refuse"-Technique

The *but you are free to accept or to refuse*-technique was introduced by Guéguen and Pascual in 2005. They were able to show that using the technique lead to increased compliance with the request of participating in a study. Guéguen and Pascual argue that the effect could be due to a semantic evocation of freedom in the subjects [14]. Since the experience of state reactance is a consequence of a perceived loss of freedom, semantically evocating freedom could be an effective way of reducing state reactance.

The stimulus texts described a smart home system that contained multiple assistant systems, such as a cooking assistant or an energy saving assistant. The systems followed different strategies, while the high-threat version explicitly restricted the user's freedom, the low-threat version did not. The stimuli have already been published in [12].

8.4.3   Procedure

All participants received an invitation email via the participant platform that was used. The email contained a link to a website where the participants could register their email address. A link to the online study was then send to that email address. The online study was realized via an online survey, using the tool Limesurvey 1.90+ [25]. The first question of the survey asked the participants to state their age. This information was used to distribute the participants to either condition. Participants that provided an even number were assigned to the condition with the high-threat stimulus, whereas participants that provided an odd number were assigned to the condition with the low-threat stimulus. This method was used to facilitate an even distribution of age over the two conditions to minimize the risk of age-related biases. The participants were then presented with either the high-threat or the low-threat stimulus and were instructed to read it carefully. After they had finished reading, they could proceed to the questionnaire items. The stimulus text was also provided on the page where the items had to be completed. The items had to be answered using a five-point Likert-type [24] scale ranging from "Strongly disagree" to "Strongly agree" in a forced-response scenario. The participants were only able to proceed to the next page if all items were completed. The next page contained five questions that were used as control questions. The control questions were used to check if the participants had really read the stimulus texts carefully. The control questions consisted of two free-text questions and three multiple-choice questions, one of which asked the participants directly if they had read the text carefully. After completing the control questions, the participants could decide if they wanted to participate in the drawing of three times 50 Euro. If they wanted to participate they had to provide their email address and their name.

8.5   Analysis

All complete datasets were reviewed to eliminate invalid datasets. Datasets were regarded as invalid if not all of the five control questions were answered correctly, which would indicate that the respective participant did not read the stimulus text thoroughly, or if all items were answered with the same response, e.g., always "Strongly agree". After eliminating the invalid ones, 342 datasets remained for further analysis. At first, the datasets were divided into two groups at random, resulting in one group of 162 datasets (A) and one group of 179 datasets (B). These were then used to perform a factor analysis following the method that was used in [33] and proposed by Homburg and Gierig [15], which is described in the following.

8.5.1   Item Selection

At first, dataset A was used to calculate item difficulty. Item difficulty is calculated in the following manner:

$$Difficulty(Item) = \frac{\sum_{i=1}^{n} X_{vi}}{n * max(X_i)} * 100 \qquad (8.1)$$

with
$\sum_{i=1}^{n} X_{vi}$ = Sum of all ratings of one Item by i subjects.
$n * max(X_i)$ = Sum of the ratings of one Item if all participants rate the maximum score [11, p. 477].

Items that show an item difficulty of 0–20% are extremely difficult items that almost nobody agrees to. Items that show an item difficulty score of 80–100% are items that almost everyone agrees to. As those items are not very informative and do not show high differences between individuals, it is recommended to remove these items [11, p. 477]. All 37 items that were used in the current study fell within the range of 20–80%. Item difficulty varied between 40.73 and 77.17, so all items were kept for further analysis.

In the next step, item discriminant indices were used to reduce the item set. All items that were formulated to represent either Anger, negative cognitions or a perceived freedom threat were grouped in one factor. The corrected item-total-correlation was measured using SPSS 23 [21]. The test shows how similar an item behaves toward other items in terms of variance. This is done by calculating a correlation coefficient [3, p. 507] between on item and the sum of all items. This is then corrected by the item's individual contribution to the total correlation [22]. Items that showed a score lower than 0.6 among the other items of their respective factor were removed from further analysis. Then, a maximum likelihood factor analysis was performed on all remaining items of each proposed factor, using varimax rotation. Further, the analysis was constrained to form only one factor, as proposed by [15].

Table 8.1 Indicees of the three factors after initial item selection

Factor	α	χ^2	df	Sig.
Anger	0.898	6.507	5	0.260
Negative congitions	0.945	15.539	14	0.342
Freedom threat	0.878	8.978	5	0.110

The analysis provides a goodness-of-fit test. If the goodness-of-fit test shows a significant result at the level of $p < 0.05$, then the null hypothesis can be rejected and the single factor solution is unlikely to fit the data. In this case, the procedure was repeated with a fixed number of two factors. Only the items that loaded on one factor that explained at least 50% of variance were kept. This procedure resulted in a highly reduced number of items for each proposed factor. Seven out of eleven items remained for factor anger. The factor negative cognitions still contained five out of eleven items and the factor freedom threat still contained six out of initially 16 items[3] Chronbach's α, explained variance and results of the χ^2-test are shown in Table 8.1. This factors composed the preliminary version of the questionnaire. These factors and their respective items were then used to further refine the questionnaire by means of structural equation modeling.

8.5.2 Structural Equation Modelling

Structural equation modeling is a confirmatory approach to test an assumed structure of constructs behind some phenomenon involving a latent variable, that can not be measured directly, such as state reactance. It uses statistical techniques to model multivariate relationships between involved constructs, such as e.g., anger, negative cognitions and perceived freedom threat. A detailed description of the involved processes is provided in Chap. 2 of Byrne's book on structural equation modeling with AMOS [6]. Structural equation modeling can be used to conduct a confirmatory factor analysis, which is described in the following. A structural equation model was created, based on dataset B, which contained the responses of 179 persons. According to Baltes-Götz, a structural equation model should be calculated with a sample that is larger than $N = 100$ [2, p. 27]. Additionally, Anderson and Gerbig claim that a sample size larger than $N = 150$ should "usually" lead to a proper solution for goodness-of-fit indicees [1]. Therefore, the currently used sample size of 179 should be adequate.

---

[3] One item could represent the factors anger and freedom threat (see Item 16 of Appendix A) and is therefore counted twice in this listing. The item was eventually removed from the final questionnaire.

The model was created using IBM SPSS Amos Version 24.0.0 [20]. The model allowed the three factors to correlate for the following reasons:

- According to the intertwined model [10], state reactance can be regarded as a construct comprised of anger and negative cognitions. As both factors are expected to show heightened values if a freedom threat is perceived, they are allowed to correlate.
- The factor freedom threat is a factor that can be used to confirm that a freedom threat is perceived by the subject. As a threat to freedom is a cause for state reactance, its two components anger and negative cognitions should be correlated with freedom threat.

Amos was used to calculate four fit indices, based on dataset B, that provide an overview of how well the proposed model represents the factor structure that was identified in Sect. 8.5.1, using dataset A. The fit indices and the respective criteria for appropriateness of the model were drawn from Iacobucci [19], Homburg and Gierig [15] and Byrne [6]. The fit indices and corresponding criteria that were used are explained in the following.

Comparative Fit Index (CFI $\geq$ 0.95)

This is a method in which the performance of the proposed model is tested against an independence or null model [6]. The CFI score of a model ranges from 0 to 1, where 1 represents the best fit. Iacobucci recommends a cut-off value of 0.95 for the CFI [19], which is in line with reports of Byrne [6] that 0.95 has recently been argued for in the scientific community.

Root Mean Square Error of Approximation (RMSEA $\leq$ 0.08)

The Root Mean Square Error of Approximation is described as one of the most informative criteria in covariance structure modeling [6]. It measures the discrepancy of the model with optimally chosen parameter values with the population covariance matrix [6]. This means that a low value means a good model fit, whereas a high value indicates a bad model fit. The values can range from 0 to 1. Iacobucci argues against using this measure because it tends to over-reject models when the population is rather small (N < 200) [18, 19], which is the case with the current dataset. However, Byrne argues for still using the metric, claiming that values between 0.08 and 0.1 indicate a mediocre fit and values below 0.06 indicate a good fit [6]. Taking the assumed tendency of over-rejecting models with small datasets into account, a cut-off value of .08 will be applied here, rather than 0.06.

Adjusted Goodness of Fit Index (AGFI $\geq$ 0.9)

The Adjusted Goodness of Fit Index is similar to the Goodness of Fit Intex (GFI). Both indices measure the relative amount of variance and covariance that is explained. AGFI differs from GFI in that it adjusts for the degrees of freedom. This means that it includes a penalty for added parameters [7, pp. 115]. Homburg and Gierig recommend keeping models that show an AGFI $\geq$ 0.9.

Adjusted χ^2 by the Degrees of Freedom (df) ($\frac{\chi^2}{df} \leq 3$)

The $\chi^2 - test$ indicates how well a proposed model fits. A higher χ^2 value indicates a better model fit. However, the χ^2 value is sensitive to the sample size. To account for this, researchers suggest to adjust the χ^2 value by the degrees of freedom (df) [6, 15]. The $\chi^2 - criteria$ that is applied here is $\frac{\chi^2}{df}$ should not be larger than three. This criterion is not derived mathematically but is rather consensus among many scholars [6, 15, 19].

The model initially did not satisfy the selected criteria and was therefore adapted iteratively until it satisfied all four criteria. This was done following modification indices that are provided by Amos. Specifically, the standardized residual covariances of the included items were used. Items that showed a high standardized residual covariance (higher than 1) with at least two other items were removed and the model was tested again. This procedure was applied until the model satisfied all previously specified criteria. The final model is depicted in Fig. 8.1. The corresponding fit indices are shown in Table 8.2.

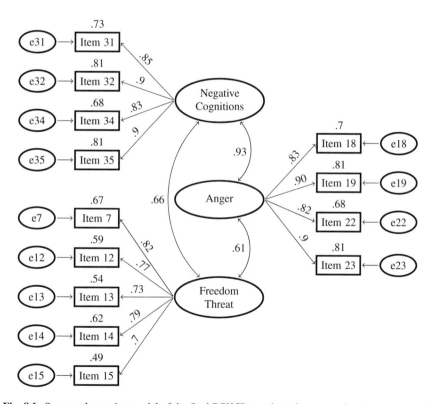

Fig. 8.1 Structural equation model of the final RSHCI questionnaire: squared nodes = measured items with R^2 item reliabilities, round nodes = RSHCI factors, rounded, double headed arrows = factor correlation with correlation weights, straight arrows = regression paths with factor loadings

Table 8.2 Fit indicees for the final structural equation model of the RSHCI questionnaire

| | | | Goodness-of-fit test | | | |
CFI	RMSEA	AGFI	χ^2	df	$\frac{\chi^2}{df}$	Sig.
0.993	0.034	0.914	74.945	62	1.209	0.125

References

1. Anderson, J.C., Gerbing, D.W.: The effect of sampling error on convergence, improper solutions, and goodness-of-fit indices for maximum likelihood confirmatory factor analysis. Psychometrika **49**(2), 155–173 (1984). https://doi.org/10.1007/BF02294170
2. Baltes-Götz, B.: Analyse von Strukturgleichungsmodellen mit Amos 18. Universität Trier (2015). Retrieved March 17, 2016 from https://www.uni-trier.de/fileadmin/urt/doku/amos/v18/amos18.pdf
3. Bortz, J., Dring, N.: Forschungsmethoden und Evaluation für Statistik für Human- und Sozialwissenschaftler, 4th edn. Springer (2006)
4. Brehm, J.W.: A Theory of Psychological Reactance. Academic Press, New York (1966)
5. Brehm, S.S., Brehm, J.W.: Psychological Reactance: A Theory of Freedom and Control. Academic Press, New York (1981)
6. Byrne, B.M.: Structural Equation Modeling with AMOS: Basic Concepts, Applications, and Programmings. Psychology Press, Tylor & Francis Group (2010)
7. Byrne, B.M.: Structural Equation Modeling with LISREL, PRELIS, and SIMPLIS: Basic Concepts, Applications, and Programmings. Psychology Press, Tylor & Francis Group (1998)
8. Dillard, J.P., Kinney, T.A., Cruz, M.G.: Influence, appraisals, and emotions in close relationships. Commun. Monogr. **63**(2), 105–130 (1996). https://doi.org/10.1080/03637759609376382
9. Dillard, J.P., Peck, E.: Affect and persuasion: emotional responses to public service announcements. Commun. Res. **27**(4), 461–495 (2000). https://doi.org/10.1177/009365000027004003
10. Dillard, J.P., Shen, L.: On the nature of reactance and its role in persuasive health communication. Commun. Monogr. **72**(2), 144–168 (2005). https://doi.org/10.1080/03637750500111815
11. Döring, N., Bortz, J.: Forschungsmethoden und Evaluation in den Sozial- und Humanwissenschaften. Springer, Berlin, Heidelberg (2016)
12. Ehrenbrink, P., Möller, S.: Development of a reactance scale for human-computer interaction. Qual. User Exp. **3**(1), 2 (2018). https://doi.org/10.1007/s41233-018-0016-y
13. Ghazali, A.S., Ham, J., Barakova, E., Markopoulos, P.: The influence of social cues in persuasive social robots on psychological reactance and compliance. Computers in Human Behavior **87**, 58–65 (2018). 10.1016/j.chb.2018.05.016. URL http://www.sciencedirect.com/science/article/pii/S0747563218302425
14. Guéguen, N., Pascual, A.: Improving the response rate to a street survey: an evaluation of the "but you are free to accept or to refuse" technique. Psychol. Record **55**(2), 297–303 (2005). https://doi.org/10.1007/BF03395511
15. Homburg, C., Giering, A.: Konzeptualisierung und operationalisierung komplexer konstrukte: ein leitfaden für die marketingforschung. Marketing: Zeitschrift für Forschung und Praxis **18**(1), 5–24 (1996)
16. Hong, S.M., Page, S.: A psychological reactance scale: Development, factor structure and reliability. Psychol. Reports **64**(3_suppl), 1323–1326 (1989). https://doi.org/10.2466/pr0.1989.64.3c.1323
17. Hong, S.M.: Hong's psychological reactance scale: a further factor analytic validation. Psychol. Reports **70**(2), 512–514 (1992). https://doi.org/10.2466/pr0.1992.70.2.512

18. Hu, L., Bentler, P.M.: Cutoff criteria for fit indexes in covariance structure analysis: conventional criteria versus new alternatives. Struct. Equ. Model.: Multidiscip. J. **6**(1), 1–55 (1999). https://doi.org/10.1080/10705519909540118

19. Iacobucci, D.: Structural equations modeling: Fit indices, sample size, and advanced topics. Journal of Consumer Psychology **20**(1), 90–98 (2010). https://doi.org/10.1016/j.jcps.2009.09.003

20. IBM Corp.: Ibm spss amos, : Wexford. Amos Development Corporation, PA (2015)

21. IBM Corp.: Ibm spss statistics for windows: Armonk. IBM Corp, NY (2015)

22. IBM Corp.: Item-total correlations in spss (2018). Retrieved May 05, 2018 from http://www-01.ibm.com/support/docview.wss?uid=swg21477000

23. Joinson, A.: Social desirability, anonymity, and internet-based questionnaires. Behav. Res. Methods Instrum. Comput. **31**(3), 433–438 (1999)

24. Likert, R.: A technique for the measurement of attitudes. Arch, Psychol (1932)

25. LimeSurvey GmbH: Limesurvey (2010). URL https://www.limesurvey.org/

26. Lindsey, L.L.M.: Anticipated guilt as behavioral motivation an examination of appeals to help unknown others through bone marrow donation. Human Commun. Res. **31**(4), 453–481 (2005)

27. Merz, J.: Fragebogen zur messung der psychologischen reaktanz. Diagnostica **29**(1), 75–82 (1983)

28. Miller, C.H., Lane, L.T., Deatrick, L.M., Young, A.M., Potts, K.A.: Psychological reactance and promotional health messages: the effects of controlling language, lexical concreteness, and the restoration of freedom. Human Commun. Res. **33**(2), 219–240 (2007). https://doi.org/10.1111/j.1468-2958.2007.00297.x

29. Quick, B.L., Stephenson, M.T.: The reactance restoration scale (rrs): a measure of direct and indirect restoration. Commun. Res. Reports **24**(2), 131–138 (2007). https://doi.org/10.1080/08824090701304840

30. Roubroeks, M., Ham, J., Midden, C.: The dominant robot: Threatening robots cause psychological reactance, especially when they have incongruent goals. In: Ploug, T., Hasle, P., Oinas-Kukkonen, H. (eds.) Persuasive Technology, pp. 174–184. Springer, Berlin Heidelberg (2010)

31. Sittenthaler, S., Traut-Mattausch, E., Steindl, C., Jonas, E.: Salzburger state reactance scale (ssr scale): Validation of a scale measuring state reactance. Zeitschrift für Psychologie **223**, 257–266 (2015). https://doi.org/10.1027/2151-2604/a000227

32. Tucker, R.K., Byers, P.Y.: Factorial validity of merz's psychological reactance scale. Psychol. Reports **61**(3), 811–815 (1987). https://doi.org/10.2466/pr0.1987.61.3.811

33. Wechsung, I.: An evaluation framework for multimodal interaction. Ph.D. thesis, Springer (2014)

Chapter 9
Validation and Intermediate Discussion

9.1 Introduction

In order to validate the questionnaire, two approaches were followed. The first approach was testing for criterion validity. Criterion validity is given if a measurement tool satisfies criteria that are in line with the theory of the investigated phenomenon [9].

The two stimuli that were used in the online study were designed according to established formulation strategy from other experiments investigating state reactance. They were designed in order to provide a high threat to freedom and a low threat to freedom (see Sect. 8.2.2 for details). The following criteria were defined for testing criterion validity:

Criterion 1 *Criterion 1 therefore states that the freedom threat dimension of the RSHCI questionnaire should show a higher score in the high-threat condition, compared to the low-threat condition.*

Criterion 1.1 *If the high-threat condition poses a high threat to freedom of the subject, according to reactance theory, this should result in state reactance. Anger is one component of state reactance. Consequently, the anger dimension of the RSHCI questionnaire should show a higher score in the high-threat condition.*

Criterion 1.2 *Such as Anger, also negative cognitions are a component of state reactance, according to the intertwined model. Following the argumentation from Criterion 1.1, also the measure of RSHCI for negative cognitions should show a higher score in the high-threat condition.*

The second approach is a special case of criterion validity [9, p. 78]. Inner criterion validity for RSHCI would be given, if a high correlation with another metric, that is

© Springer Nature Switzerland AG 2020
P. Ehrenbrink, *The Role of Psychological Reactance in Human–Computer Interaction*, T-Labs Series in Telecommunication Services,
https://doi.org/10.1007/978-3-030-30310-5_9

assumed to be valid, can be shown. For this reason, the RSHCI questionnaire was going to be used alongside the state reactance assessment method of Dillard and Shen [2] in a laboratory experiment.

Criterion 2 *If inner criterion validity is given, then state reactance, assessed with the RSHCI questionnaire should correlate highly with state reactance, assessed with the method introduced by Dillard and Shen [2].*

9.2 Smart Home Online Study Results

The results from the online experiment were calculated with the newly derived questionnaire. The descriptive results for the two described smart homes are shown in Table 9.1.

The results were analyzed with independent samples t-tests. A Levine test for equality of variances was performed for all three factors of the RSHCI. The tests showed significant differences in variance between the two conditions in the factors of anger and negative cognitions with $p < 0.05$. Therefore, the degrees of freedom for these factors were corrected with the Welch-Satterthwaite method for further analysis with independent samples t-tests. The results of the independent samples t-tests are shown in Table 9.2.

Table 9.1 Descriptive results of the RSHCI questionnaire in the two conditions of the online study

	N	Anger		Negative cognitions		Freedom threat	
		Average	Std	Average	Std	Average	Std
High threat	175	2.547	1.087	2.481	1.127	3.897	0.898
Low threat	167	1.8	0.83	1.837	0.97	3.471	0.895

Table 9.2 Results of the independent samples t-tests from the factors anger, negative cognitions, and freedom threat

	df (adjusted df)	F	p
Anger	(324.651)	7.171	<0.001
Negative cognitions	(336.461)	5.676	<0.001
Freedom threat	340	4.384	<0.001

9.3   Validation Study: Intelligent Personal Assistant Study

The previously described process lead to a questionnaire that satisfied all model-fit criteria of Sect. 8.3.2 on a subset of the data from the online study. In order to further assess the consistency and validity of the questionnaire, another study was conducted. The study was a laboratory experiment in the frame of a master's thesis. A brief description of the study was already published in a paper [3].

In order to investigate if the RSHCI questionnaire measures state reactance, a laboratory experiment was conducted. In the experiment, participants interacted with three intelligent personal assistants. After the interaction, their state reactance levels were measured with the method proposed by Dillard and Shen [2]. Also, the participants completed the RSHCI questionnaire after each interaction. The results of the two measures were later compared to draw conclusions about whether they both measure the same construct.

9.3.1   Participants

Twenty four participants participated in the laboratory experiment. The participants were of an average age of 25.29 years (SD $= 5.069$), 13 of them were females. All participants were between 18 and 35 years of age and their mother tongue was German.

9.3.1.1   Procedure

After arrival, every participant completed a form of consent, a questionnaire for demographic information and the refined version of Hong's Psychological Reactance Scale [5–7]. Afterward, they received a short training on how to operate each of the three intelligent personal assistants (Siri [1], Cortana [8], Google Now [4]), which were installed on three smartphones, via voice commands. Afterward, the main experiment would start. The experiment consisted of three conditions. In each condition, the participants were asked to complete a series of tasks with one of the three assistants. To make the conditions comparable, tasks were chosen that could be performed by each of the three assistants. Even though all assistants are somewhat robust in their understanding of basic request, prior knowledge of the voice commands is helpful. Teaching all those commands could not be accomplished in the limited timeframe of a laboratory experiment. Therefore, all participants received a sheet of paper on which the tasks and the appropriate commands for the tasks were written down.

The order of condition was systematically changed so that every possible order was completed by four subjects. This procedure ensured that sequence-effects would not affect the results in favor of a particular condition. Each condition covered five sets of tasks that consisted of several sub-tasks each. All tasks required a voice command by the subject. In total, the participants had to state 24 voice commands in each condition. As the appropriate voice commands were provided, the command-error rate was not accessed. However, the examiner reported that it has been very low. After each condition, the participants were asked to fill out the RSHCI and the anger questionnaire [2]. Afterward, they conducted the thought-listing task together with the examiner.

9.3.1.2   Results

The RSHCI questionnaire consists of the three dimensions anger, negative cognitions, and freedom threat. The freedom threat dimension can be regarded as a confirmatory factor and is therefore not necessary for assessing state reactance, as such. Internal consistency of the two RSHCI factors anger and negative cognitions was high with Cronbach's $\alpha = 0.897$ for anger and $\alpha = 0.842$ for negative cognitions.

To correlate the state reactance measures, a state reactance score was calculated for the RSHCI by averaging all items that belong to the factors anger and negative cognitions. For the method of Dillard and Shen, the score was calculated by averaging all items of the anger questionnaire. This score was then z-transformed. Also, the number of negative cognitions to the thought-listing task was z-transformed. Then, the two z-transformed scores were added to a single state reactance score. This state reactance score of the method by Dillard and Shen was correlated to the state reactance score of the RSHCI. The results show a highly significant, high, positive correlation between the two-state reactance measures with $r = 0.807$, $n = 72$, $p < 0.001$. Also, a highly significant but rather low correlation was observed between the refined version of Hong's Psychological Reactance Scale and the RSHCI, with $r = 0.323$, $n = 72$, $p < 0.01$. An overview of the results that are used for validation can be seen in Table 9.3.

Also, correlations between the two factors anger and negative cognitions of state reactance measurements were correlated. The correlation between the two measures of anger was high, positive, and highly significant with $r = 0.821$, $n = 72$, $p < 0.001$.

Table 9.3 Correlation of the state reactance measures Reactance Scale for Human–Computer Interaction (RSHCI) and the method introduced by Dillard and Shen, as well as the trait reactance measure of the HPRS (revised version). Highly significant correlations are marked with doubles asterisk

	RSHCI	Dillard Shen
HPRS	0.323**	0.213
RSHCI		0.807**

The correlation between the two measures of negative cognitions was of medium height and highly significant with r = 0.470, n = 72, p < 0.001.

9.4 Intermediate Discussion on the Reactance Scale for Human–Computer Interaction

The thoroughly conducted development process of the RSHCI that is described in Chap. 3 has resulted in a well-constructed questionnaire. It surpassed all criteria that have been set for model-fit indices and can be used in a long, 13-item version to measure perceived freedom threat and state reactance via its components anger and negative cognitions, or in a short, eight-item version to measure state reactance via its components, only.

9.5 Methodological Considerations

Hong's Psychological Reactance Scale

In the study of Chap. 6, the HPRS was used in a reduced version which was proposed by Yost [10]. In the experiment described in Sect. 9.3, another version was used, namely the refined version [5]. The refined version was used because also Dillard and Shen used this version to validate their measurement technique. To achieve better comparability, it was decided to do the same.

Handing out Correct Commands to the Participants

The participants of the intelligent personal assistant study have received a sheet of paper where the correct voice commands for all tasks were listed for each intelligent personal assistant. This procedure introduced a bias to the experiment that prohibits interpretation of intuitiveness of the commands for the intelligent personal assistants. However, this procedure also allowed for the experiment to remain relatively unobstructed by problems regarding missing expertise of the users with the intelligent personal assistants. One goal (that was not related to the current work) of the experiment was to gather data for recommending intelligent personal assistants to expert users. Not handing out the correct commands would have probably introduced more variance into the state reactance data and would have therefore resulted in more reliable results.

Discriminant Validity

Figure 8.1 shows that the two factors anger and negative cognitions are highly correlated, suggesting that they could be regarded as a single factor for state reactance. This is also indicated by a one-factorial solution if the corresponding items are factor-analyzed with varimax rotation. However, the intertwined process cognitive-affective model, which has been shown to be the best fitting model for state reactance dictates two separate factors, which, as being psychological constructs can also be correlated. In order to allow for this correlation, the factor analysis was again performed, using Promax rotation, with revealed the two factors (pattern matrix). The structural equation model was built regarding anger and negative cognitions as two correlated factors. The Fornell-Larcker-criterion ($AVE_X > R^2_{X,Y}$), which tests discriminant validity is met if anger and negative cognitions are regarded as a single factor for state reactance, however, it is not met, if they both are implemented as two factors. Despite not meeting the Fornell-Larcker-criterion, the model was kept regarding anger and negative cognitions as two separate factors, following the theoretical assumptions. Further support for the structure of the model fitting the data is also coming from χ^2-criterion, which was passed.

State Reactance Score

A single score for state reactance was calculated from the factors anger and negative cognitions of the RSHCI questionnaire. The questionnaire also contains a dimension for assessing a perceived threat to freedom. One could argue that this dimension should also be included in the total score of state reactance. The RSHCI was developed following the intertwined model of state reactance [2], thereby assuming anger and negative cognitions as components of state reactance. The freedom threat part of the questionnaire is intended to be a confirmatory variable for state reactance assessment and was therefore not used for a total state reactance score. However, it could be argued that freedom threat is a prerequisite for state reactance to appear and should, therefore, be included in the total score. Such a method of analysis would still be a valid one. However, in the line of this work, it was decided to follow the intertwined model to allow for easier interpretation of the results along adjacent research.

9.6   Validation

In order to investigate criterion validity of the RSHCI, several criterias were formulated that followed from psychological reactance theory. Criterion 1 regarded the freedom threat dimension of the questionnaire. Since the two stimuli of the online study were designed according to established techniques that can induce or inhibit freedom threats, the freedom threat dimension should show a difference in ratings of the two stimuli. The high freedom threat stimulus should have resulted in higher scores in freedom threat, compared to the low freedom threat stimulus. The obtained results of the online study are shown in Tables 9.1 and 9.2. The scores of freedom threat were significantly higher for the high-threat stimulus, compared to the low-threat stimulus. Criterion 1 is therefore satisfied.

Criterions 1.1 and 1.2 stated, that if the high-threat stimulus really poses a higher threat to freedom than this should result in state reactance. If the participants experience state reactance, and the RSHCI can measure its components, Anger, and negative cognitions, then the scores for the RSHCI factors anger and negative cognitions should be higher for the high-threat stimulus. Indeed, as Tables 9.1 and 9.2

show, the scores for anger and for negative cognitions were significantly higher for the high-threat condition than for the low-threat condition. Hence, Criterions 1.1 and 1.2 can be regarded as satisfied.

Criterion 2 regarded the inner criterion validity of the scale. The inner criterion validity test assumes, that the validity of RSHCI is given if it correlates highly with a measure of which validity can be assumed (such as the state reactance measurement of Dillard and Shen [2]). The results of a calculated correlation between the total state reactance score of the method of Dillard and Shen and the RSHCI, provided in Table 9.3 showed that there is a highly significant, high, positive correlation between the two measures, suggesting that inner criterion validity is given and that this criterion is satisfied. Also, the two components of state reactance were compared. The two anger measurements also showed a highly significant, high, positive correlation. However, the negative cognitions component only showed a medium positive correlation, which was also highly significant. One explanation for this could be that the measurement of Dillard and Shen, which is the number of negative cognitions in a thought-listing task, is completely free, meaning that the participants can say whatever they want. In the RSHCI questionnaire, the answers are pre-formulated and the participants can only answer via different degrees of agreement. Even though the pre-formulated answers are formulated quite broadly, it is possible, that they do not cover all aspects of negative thoughts. This is probably a shortcoming of the RSHCI questionnaire. But the correlation between the two negative cognitions measures is of medium height. A criticism of the thought-listing technique was that it could be vulnerable to a social desirability bias. Such a bias is not as strong in the RSHCI questionnaire since it does not involve direct human–human interaction. It could therefore also be the case that the RSHCI provides the more accurate measure of state reactance, evidence for this is the highly significant, although low, correlation of the RSHCI total score with trait reactance (see Table 9.3), which is not present for the measure of Dillard and Shen.

References

1. Apple Inc.: Siri (2011). Retrieved December 23, 2016 from http://www.apple.com/ios/siri/
2. Dillard, J.P., Shen, L.: On the nature of reactance and its role in persuasive health communication. Commun. Monogr. **72**(2), 144–168 (2005). https://doi.org/10.1080/03637750500111815
3. Ehrenbrink, P., Osman, S., Möller, S.: Google Now is for the Extraverted, Cortana for the introverted: Investigating the influence of personality on IPA preference. In: Proceedings of the 29th Australian Conference on Human-Computer Interaction, pp. 1–9. ACM, New York, NY (2017). https://doi.org/10.1145/3152771.3152799. Electronic, online
4. Google Inc.: Google now (2012). Retrieved December 23, 2016 from https://www.google.com/landing/now/
5. Hong, S.M., Faedda, S.: Refinement of the hong psychological reactance scale. Educ. Psychol. Measurement **56**(1), 173–182 (1996). https://doi.org/10.1177/0013164496056001014

6. Hong, S.M., Page, S.: A psychological reactance scale: Development, factor structure and reliability. Psychol. Reports **64**(3_suppl), 1323–1326 (1989). https://doi.org/10.2466/pr0.1989.64.3c.1323
7. Hong, S.M.: Hong's psychological reactance scale: a further factor analytic validation. Psychol. Reports **70**(2), 512–514 (1992). https://doi.org/10.2466/pr0.1992.70.2.512
8. Microsoft: Cortana (2014). Retrieved December 23, 2016 from https://www.microsoft.com/windows/cortana/
9. Sedlmeier, P., Renkewitz, F.: Forschungsmethoden und Statistik in der Psychologie. Pearson Studium (2008)
10. Yost, A.B., Behrend, T.S., Howardson, G., Badger Darrow, J., Jensen, J.M.: Reactance to electronic surveillance: a test of antecedents and outcomes. J. Bus. Psychol. (2018). https://doi.org/10.1007/s10869-018-9532-2

Part IV
Determinants of State Reactance

Chapter 10
Smart TV Study—System Errors

10.1 Research Question: What Factors Influence State Reactance?

Chapter 5 identified several situations in which state reactance might be triggered by interaction with technical systems. One of these situations that has not been investigated in the literature so far, is the occurrence of system errors. Besides the situations that have been identified in Chap. 5, Sect. 2.2.3 has introduced possible moderator variables for state reactance. These moderators have not yet been investigated in the context of human–computer interaction. The moderators that have been mentioned are trait reactance, social agency, and involvement of the user (Sect. 2.2.3). For human–computer interaction, this factors could be of relevance if they have an influence on state reactance and its potential consequences.

Research Question 3 *Factors that influence state reactance can be moderator variables, which indirectly influence interaction with a technical system by changing the effect that state reactance has on the interaction, or they can directly influence state reactance, e.g., by triggering it. Research Question 3 asks: What factors influence state reactance?*

Research Question 3.1 *Several moderator variables for state reactance have been described that might also play a role in human–computer interaction. Research Questions 3.1 therefore asks: Are trait reactance, social agency, and involvement moderator variables for state reactance in the context of human–computer interaction?*

Research Question 3.2 *If factors can be identified that influence the level of state reactance or the effect that state reactance has on interaction with technical systems, then these factors might be used to reduce negative consequences of state reactance. Research Question 3.2 therefore asks: Can the factors that influence state reactance be used to reduce effects of state reactance on the technical system.*

© Springer Nature Switzerland AG 2020
P. Ehrenbrink, *The Role of Psychological Reactance in Human–Computer Interaction*, T-Labs Series in Telecommunication Services,
https://doi.org/10.1007/978-3-030-30310-5_10

10.2 Introduction

The expert survey of Chap. 5 identified a number of situation in human–computer interaction in which the participants reported to have experienced state reactance. All but one of these situations were covered by the literature that was analyzed in Chap. 4. The one situation that was missing in human–computer interaction literature was encountering system errors. It was also the second most often reported situation after reduced freedom of choice, which indicates that system errors are a relevant aspect of the role of psychological reactance in human–computer interaction and that a knowledge gap exists concerning system errors. Therefore, system errors were investigated in the laboratory experiment that is described in this chapter.

The experiment has been conducted during the development of the RSHCI questionnaire. Measurement of state reactance has been accomplished with an early set of items that were later used for the development of the final questionnaire. This set of items did contain all items that were later used for the dimensions anger and negative cognitions. Hence, state reactance is going to be operationalized with these items. However, the item set did not yet contain the items for the freedom threat dimension, freedom threat was therefore not assessed. Results of this study have been published in [6]. The study has been designed and analyzed by the author. However, the analysis in the paper correlated state reactance and acceptability independently for the two subject-groups, whereas the correlation was calculated over both groups in the analysis presented here. Since the experiment has been published by the author, also texts in this chapter are derived from that publication.

Developers usually try to design devices in a way that system errors are avoided as much as possible. Still, sometimes, system errors do happen. In the expert survey, the reports about system errors usually involved a description, that the error either prevented or at least delayed the participants from reaching their goal.

Hypothesis 4 *System errors are undesired events that occur when users interact with a technical system. Participants in Chap. 5 reported that system errors triggered state reactance. Hypothesis 4 therefore states that system errors cause state reactance.*

As explained in Chap. 2, the trigger of reactance is usually the perceived loss of personal freedom or control [2, 3]. If a user interacts with a device, a system error is most often an undesired event and therefore constitutes a temporary loss of control over the device or the situation. Since system errors can often not be avoided completely, error handling has to be implemented in a way that it reduces the negative effects of the errors as much as possible. Assuming that system errors do cause a threat to the perception of the user of being in control, state reactance could be avoided by reestablishing the lost freedom. If was argued in Chap. 1, that users can fall back on interaction schemes from human–human interaction more easily, if they cannot comprehend the functionality of the system they interact with. The same argument could be used for system errors. If users cannot comprehend why a system produces the error, they might fall back on state reactance. But, if users would be more aware of the reason why an error occurred, or even the chain of events that led to the error, they might be more confident about resolving the error. Also, they might be able to avoid it completely, next time. Knowing the reason for a current system error increases situational awareness for the user while interacting with a system, which in turn might reestablish their perceived control over the situation after a system error.

Hypothesis 5 *The trigger of state reactance is usually a perceived loss of control. This effect might be increased if the participants cope with the situation by applying schemes from human–human interaction, because they do not rationally understand why the error has happened. If users are aware of the reasons that led to an error, they might feel more confident about resolving or avoiding the error. Therefore, more awareness over the reasons for an error will result in less state reactance.*

The study was also used for the purpose of validating the RSHCI questionnaire. Psychological state reactance, as it was intended to be measured by the RSHCI can result in a downgrading of the perceived freedom threat and thereby reduce acceptance of the source of that threat [3, 5]. Directly accessing acceptance of a device that is not yet available for customers, like the smart TV in the present study, is not possible, because the real buying behavior or long-term satisfaction with the product cannot be accessed in the frame of a laboratory experiment. However, one can draw conclusions about a devices acceptance-potential by measuring its acceptability. The smart TV was being controlled via voice commands and also gave feedback via voice output. Interaction could involve multiple turns so that the smart TV could be regarded as a spoken dialogue system. For spoken dialogue systems like the smart TV, acceptability can be measured by the C1 component of the P. 851 questionnaire [8, 9]. The P. 851 questionnaire was part of the study so that a correlation could be calculated between acceptability and state reactance, as it is measured by the RSHCI. This also allowed for conclusions about the criterion validity of the RSHCI.

Hypothesis 6 *It was stated in Sect. 2.2.1.1, that source derogation is an indicator of state reactance. Source derogation is a downgrading of the source of the perceived threat to freedom or control. So, if a user perceives a threat to his or her control over a device, this might downgrade the user's opinion of that device, resulting in less acceptance. Acceptance is difficult to measure, but acceptability is the potential for a device to be accepted by users. If a technical system causes state reactance, its acceptance by the user can decline, this will be indicated by a decrease of acceptability. State reactance should, therefore, correlate negatively with acceptability.*

10.3 Participants

36 adult participants took part in the experiment. Their age varied between 18 and 44 years with a mean age of 27.44 years (SD = 5.60). Gender was balanced among the group of subjects, resulting in a total of 18 males and 18 females. For the analysis, the dataset of one participant was dropped, as it represented an extreme outlier with a value of the factor anger, that differed more than three times the standard deviation from the average population's value of anger.

10.4   Methods

A modified version of a smart TV interface from the company LOEWE., that was used
in the BMBF-Project Universal Home Control Interfaces @ Connected Usability
(UHCI) served as a use case. The interface was composed of two parts on two devices.
The input part was a smartphone app, that was running on an Android phone. The
Android app was a voice-operated remote control for the smart TV. It only contained a
push-to-talk button as the only interface-element that the participants were supposed
to trigger. After pushing the button, the participants could give a voice-command
that the smart TV would then execute. The output part was the interface of the
smart TV. It contained three fields that provided feedback over the processing of
the voice command (see Fig. 10.1), which will be described in more detail in the
Sect. 10.4.1. The experiment was a partial within-partial between-participants design
with three conditions. The baseline condition was completed by all subjects, the two
other conditions were completed by half of the participants respectively. The three
conditions are illustrated in detail in Sect. 10.4.1.

10.4.1   Feedback Types

The partial within-partial between-participants design allowed for testing three conditions on two groups of subjects. Before going into further detail of the conditions,
the three feedback types of the Smart TV interface need to be explained. Fig. 10.1
shows a photo of the interface as it appeared on the TV. The TV is operated via
natural-language style voice commands that are interpreted and then executed. The
TV gives three types of feedback after a speech-command is perceived.

Feedback A

Feedback A was written feedback about the speech command as it was recognized by
the speech-to-text system. Participants could give commands in natural language and
they would then appear as text on the upper left of the smart TV interface. Feedback
A, therefore, provided feedback about whether the speech command was recognized
successfully.

Feedback B

Feedback B provided feedback about the interpretation of the speech command after
it was recognized by the speech-to-text system. Participants could see in the upper
right part of the screen if the smart TV interpreted the speech command that they
gave or that the speech-to-text system recognized correctly or not.

Feedback C

Feedback C was the main area on the screen that provided output. After the smart
TV interpreted a command, it would display the resulting function, e.g., a YouTube

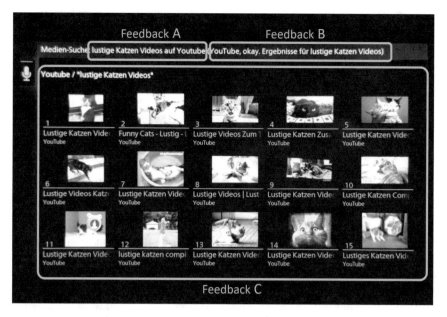

Fig. 10.1 Feedback A: speech to text output; Feedback B: Interpretation of System; Feedback C: System output

search result or a video, in this area. The participants were able to see via feedback C if the smart TV executed a command correctly or not.

10.4.2 Conditions

The experiment was conducted using three conditions. One condition was a baseline condition. It did not include any artificially introduced errors. After a participant gave a speech command, the recognized word would appear in feedback A, the interpreted command would appear in feedback B and the command would be executed in feedback C. The processing chain of a voice command in this condition can be seen in Fig. 10.2.

The second condition contained artificially introduced interpretation errors. The errors were introduced in the interpretation stage of the processing chain. Voice com-

Fig. 10.2 Baseline condition of the Smart TV study. No artificial errors were introduced in the processing chain of voice commands

Fig. 10.3 Interpretation error condition of the Smart TV study. Artificial errors were introduced in the interpretation level of the processing chain of voice commands

Fig. 10.4 Unexplained error condition of the Smart TV study. Artificial errors were introduced in the execution level of the processing chain of voice commands

mands would be recognized correctly by the speech-to-text unit but then interpreted wrongly. For example, the voice command "Show me movie trailers on YouTube" would be recognized correctly as "show me movie trailers on YouTube" but then interpreted wrongly as "(YouTube, okay: Results for unboxing videos)". The execution of the command would then follow the interpretation and feedback C would show the results of "unboxing videos", instead of "movie trailers", as the participant asked for. This condition results in false actions of the smart TV, but the participants are provided with an explanation for the error via feedback B, namely a misinterpretation of the command. The processing chain of a voice command in this condition can be seen in Fig. 10.3.

The third condition introduced artificial errors, as well. However, in this condition, the errors were introduced only at the execution level. Following the exemplary voice command "Show me movie trailers on YouTube", the smart TV would again display the correct speech-to-text recognition "show me movie trailers on YouTube". Also, it would show a correct interpretation of the command in feedback B. "(YouTube, okay: Results for movie trailers)". Only the execution of the command was erroneous in this condition. Therefore, feedback C would show results for "unboxing videos", instead of results for "movie trailers". In contrast to the other condition that contains errors, this condition does not provide an explanation of why the error occurred through feedback. The processing chain of a voice command in this condition can be seen in Fig. 10.4.

10.4.3 Introduction of Errors

The errors were introduced to the two error-conditions under three constraints:

1. No errors are introduced in the first two tasks.
2. The overall error rate is targeted at 50%.
3. No errors are introduced in the last three tasks.

These constraints were set up because the participants were supposed to regard the errors as real system errors, if the rate would have been too high, participants would probably have noticed the artificial nature of the errors, especially those, who had completed the baseline condition before. On the other hand, the error rate needed to be high enough to be salient over the whole set of tasks.

Also, the first two tasks and the last three tasks were excluded from the list of tasks where the errors could appear, this was to avoid primacy and recency effects that could have resulted in the participants deeming the Smart TV as completely malfunctioning.

Primacy Effect

The primacy effect is a memory effect that results in persons better memorize items that are presented early on in a list of items because these can be integrated more easily into long-term memory [11, p. 924]. The effect is of importance in test situations because the list of tasks that the participants have to complete is also prone to primacy effects, which could result in a participant more strongly memorizing their experience with the first tasks, which then influences their final judgment of the system under investigation [4].

Recency Effect

The recency effect is similar to the primacy effect but results in a better ability to memorize items that are presented in the end of a list of items because these items are still in the short term memory [4].

10.4.4 Experimental Procedure

After arrival, every participant was greeted and introduced to the study via an introduction sheet. The participants were informed that they could ask questions or abort the experiment whenever they liked or felt uncomfortable. After the participants had read the instruction sheet, they were handed a form of consent. After they had signed it, they were seated in front of the smart TV. The experimenter then turned on the smart TV and started the remote-control application on the smartphone. As soon as the devices were running and ready, the experimenter gave the participant a short demonstration of how the system worked and explained the purpose and the content of the three feedback fields. The participants then went through a short training in which they could try out the smart TV with a couple of queries that they had to come up with by themselves. Afterward, the experimenter asked the participants if they had any questions. If they did not have any questions, the experiment would start.

Every participant interacted with the smart TV in the baseline condition and either one of the two conditions with artificially introduced errors. The order of the conditions was changed after every participant to avoid sequencing effects. Each condition contained 17 tasks that the participants had to try out. If they failed to

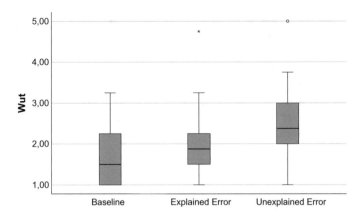

Fig. 10.5 Two outliers. The extreme outlier is represented by a star, the outlier is represented by a circle

accomplish the task after two trials, they would proceed to the next one. The tasks can be reviewed in Table.

To make sure that all participants were aware of the content of the three feedback types, they had to write down the content of the three feedback fields after every task.

After completion of all tasks of one condition, the participants were asked to fill in a couple of questionnaires on a laptop, including the RSHCI to measure state reactance and the P. 851 [9] which measures a variety of quality factors for spoken dialogue systems, including acceptability.

At the end of the experiment, the participants were asked to fill in the demographic questionnaire. It was positioned at the end of the experiment to avoid gender effects, which are known to bias experimental results if participants are asked about their personal details (including gender) prior to the experiment.

10.5   Results

The gathered data were analyzed using IBM SPSS 25 [7]. In a first step, the factor values of interest were calculated by averaging the respective items of the questionnaires. The factors were anger, negative congitions of the RSHCI and acceptability of the P. 851. Anger and negative cognitions were averaged in order to calculate a score for state reactance. The average values and the standard deviation of those factors over the three conditions can be reviewed in Table 10.1.

A subsequent visual inspection of the data, using boxplots revealed an extreme outlier in the factor anger for the condition where an explanation of the errors is provided (see Fig. 10.5).

SPSS uses the Tukey Fences [12] criterion to classify outliers, meaning that data points that are below or above the third quantile (Q_3) are regarded as *far out* out-

Table 10.1 Descriptives Statistics of the factors anger (RSHCI), negative cognitions (RSHCI), state reactance (average of negative cognitions and Anger) and acceptability (P. 851) after the removal of the extreme outlier

Condition	Anger		Negative cognitions		State Reactance		Acceptability	
	Average	STD	Average	STD	Average	STD	Average	STD
Baseline	1.71	0.72	1.83	0.77	1.77	0.67	3.55	0.70
Error Expl.	1.93	0.62	1.90	0.63	1.91	0.62	3.21	0.55
Error Unexpl.	2.56	1.02	2.64	1.04	2.60	1.02	2.70	0.78

liers [12, p. 11]. Outliers can negatively influence statistical analysis and should be removed from the dataset [10, p. 104]. After removal of the outlier, the average values and standard deviation of the four factors were recalculated and can be viewed in Tab 10.1.

In the next step, the effects of the error-type on the four factors were investigated. Because of the partial within-partial between-participants design, that include dependent and independent participant groups, an ANOVA could not be applied. Therefore, a number of pairwise comparisons were calculated to test for differences between the groups. The average values and standard deviations of each factor can be viewed in Table 10.1.

Baseline—Baseline

As a result of the experiment design, the Baseline condition consists of two independent subject-groups. To test if those groups differ significantly in the measured factors, they were statistically analyzed further. First, a Levine test for equality of variances was performed on the measured factors. This was done to find out if the groups showed equal variances and could thereby be analyzed by independent samples t-tests without correction. The results were not significant for the factors Anger, negative cognitions, and state reactance with $p > 0.05$. acceptability showed significant differences in variances with $p = 0.044$. Following this results, Anger, negative cognitions, and state reactance were analyzed with independent samples t-tests, acceptability was analyzed with independent samples t-tests after the degrees of freedom were corrected with the Welch-Satterthwaite method. Results of the independent samples t-tests showed no significant differences between the two groups in the factor anger with $t(34) = -0.57$, $p = 0.572$. Also, no significant differences between the two groups in the factor negative cognitions were observed with $t(34) = -1.20$, $p = 0.236$. State reactance also showed no significant differences between the two groups with $t(34) = -0.99$, $p = 0.328$. An independent samples t-test with adjusted degrees of freedom for the factor acceptability between the two groups indicated that there are no significant differences with $t(31.318) = 1.105$, $p = 0.278$.

Baseline—Explained Error

Paired samples t-tests were used to test for differences between the baseline and the condition with the explained error in the factors Anger, negative cognitions, state reactance, and acceptability. The test for anger showed a significantly higher score for the explained error condition than the baseline condition with $t(16) = -3{,}588$, $p < 0.01$. The test for negative cognitions showed a significantly higher score for the condition with the explained error with $t(16) = -2.176$, $p < 0.05$. State reactance also had a significantly higher score for the condition with the explained error with $t(16) = -3.125$, $p < .01$. Also, acceptability had a significantly lower rating in the condition with the explained error with $t(16) = 4.518$, $p < 0.001$.

Baseline—Unexplained Error

Paired samples t-tests were used to test if the observed differences between the baseline and the condition with the unexplained error in the factors Anger, negative cognitions, state reactance, and acceptability are significant. The paired samples t-test for anger showed a significantly higher score for the unexplained error condition than the baseline condition with $t(17) = -3{,}053$, $p < 0.01$. The test for *negative congitions* showed a singificantly higher higher score for the condition with the explained error with $t(17) = -2.536$, $p < 0.05$. state reactance also had a significantly higher score for the condition with the explained error with $t(17) = -2.950$, $p < 0.01$. Also, acceptability had a significantly lower rating in the condition with the explained error with $t(17) = 3.733$, $p < 0.01$.

Explained Error—Unexplained Error

The Levine test for equality of variances showed non-significant results for the factors Anger, negative cognitions, and acceptability with $p > 0.05$. Therefore, independent samples t-tests were performed to compare the two error conditions based on these factors. The factors anger showed a significantly higher rating in the condition with the unexplained error with $t(33) = -2.197$, $p < 0.05$. Also, negative cognitions was significantly higher in the condition with the unexplained error with $t(33) = -2.537$, $p < .05$. acceptability on the other hand was significantly higher in the condition with the explained error with $t(33) = 2.229$, $p < 0.05$. The Levine test for equality of variances was significant for the factor state reactance with $p < 0.05$, meaning that the degrees of freedom had to be adjusted. Results of the independent samples t-test with adjusted degrees of freedom show a significant difference in state reactance with higher state reactance in the conditions with the unexplained errors with $t(25{,}874) = -2.599$, $p < 0.05$.

Correlation State Reactance—Acceptability

The measurements of state reactance and acceptability from all conditions were used to investigate the relationship between the two variables. The Shapiro-Wilk test indicated that state reactance and acceptability were not normally distributed. Therefore, a Spearman's Rank-Order Correlation was calculated for state reactance and acceptability. Spearman's Rank-Order Correlation can be used to measure correlations of variables that are not normally distributed. Data is first transformed into

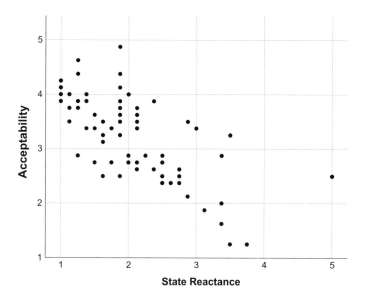

Fig. 10.6 Scatterplot showing the relationship of state reactance and *acceptabilty*

ranks and the monotonous association between the two variables is then accessed [1, pp. 232]. There was a highly significant negative correlation between state reactance and acceptability with $\rho = -0.695$, p < .001. The scatterplot in Fig. 10.6 gives an overview of the relationship between state reactance and acceptability.

References

1. Bortz, J.: Statistik für Human- und Sozialwissenschaftler. Springer (2004)
2. Brehm, J.W.: A Theory of Psychological Reactance. Academic Press, New York (1966)
3. Brehm, S.S., Brehm, J.W.: Psychological Reactance: A Theory of Freedom and Control. Academic Press, New York (1981)
4. Caplan, B., Kreutzer, J.S., DeLuca, J.: Encyclopedia of Clinical Neuropsychology; With 199 Figures and 139 Tables. Springer (2011)
5. Dillard, J.P., Shen, L.: On the nature of reactance and its role in persuasive health communication. Commun. Monogr. **72**(2), 144–168 (2005). https://doi.org/10.1080/03637750500111815
6. Ehrenbrink, P., Gong, X.G., Möller, S.: Implications of different feedback types on error perception and psychological reactance. In: Proceedings of the 28th Australian Conference on Computer-Human Interaction, OzCHI '16, pp. 358–362. ACM, New York, NY, USA (2016). https://doi.org/10.1145/3010915.3010994
7. IBM Corp.: Ibm spss Statistics for Windows: Armonk. IBM Corp, NY (2015)
8. International Telecommunication Union: Subjective quality evaluation of telephone services based on spoken dialogue systems. Tech. Rep. Supplement 851 to P-Series Recommendations, International Telecommunication Union, Geneva, Switzerland (2003)

9. Möller, S., Smeele, P., Boland, H., Krebber, J.: Evaluating spoken dialogue systems according to de-facto standards: A case study. Comput. Speech Lang. **21**, 26–53 (2007)
10. Schäfer, T.: Statistik I: Deskriptive und Explorative Datenanalyse, vol. 1. VS Verlag fr Sozialwissenschaften (2010). https://doi.org/10.1007/978-3-531-92446-5
11. Stadtler, T.: Lexikon der Psychologie. Alfred Kroner Verlag (2013)
12. Tukey, J.W.: Exploratory data analysis **1**, (1977)

Chapter 11
Persuasive Assistant Study—Moderator Variables

11.1 Introduction

The previously described studies have already been showing that there are several situations in which state reactance can be triggered in human–computer interaction. Also, it was shown that state reactance can have negative consequences for the acceptability of a technical system. This suggests, that state reactance could be used as a predictor for acceptance. However, usability evaluation and user experience research already know a variety of metrics that are used to estimate how well a technical system might be accepted. One widely used metric for this is the AttrakDiff questionnaire [6], it contains the dimensions hedonic quality, pragmatic quality, and *attractiveness*. Attractiveness can be regarded as a generalized judgment of the quality of the technical system under evaluation [6, p. 14]. According to Dieffenbach and Hassenzahl, the dimensions pragmatic quality and hedonic quality address aspects of usability and user experience, which both contribute to attractiveness to roughly the same amount [6, p. 14]. If the assessment of state reactance increases the prediction accuracy of attractiveness,[1] it could be argued that state reactance addresses an additional aspect of human–computer interaction, which can be regarded as beneficial for evaluating technical systems, and which is not covered by either hedonic quality or pragmatic quality.

Hypothesis 7 *It was shown in Sect. 9.3.1.2, that state reactance is highly correlated with acceptability. In this study, attractiveness is regarded as the potential of a system to gain acceptance. Since state reactance is correlated with acceptability, it should also be beneficial in predicting the attractiveness of a system. Hypothesis 7 claims that state reactance can be used to increase the prediction accuracy of attractiveness alongside hedonic quality and pragmatic quality.*

---

[1]Since attractiveness is a global judgment of a system, it can be regarded as an operationalization of potential acceptance.

© Springer Nature Switzerland AG 2020
P. Ehrenbrink, *The Role of Psychological Reactance in Human–Computer Interaction*, T-Labs Series in Telecommunication Services,
https://doi.org/10.1007/978-3-030-30310-5_11

Another purpose of the current experiment is to investigate the effects of the proposed moderator variables on state reactance. One of these potential moderators is trait reactance. Quick and Stephens [17] found that trait reactance was a predictor of anger. This means that there was a significant correlation between trait reactance and Anger, suggesting that trait reactance could act as a moderator variable for anger. It seems that state reactance moderators do not necessarily act on the construct of state reactance as a whole, but may also influence only one of its components. Therefore, moderating effects will be tested on the two components anger and negative cognitions separately.

Hypothesis 8 *It is currently unknown if these moderators are also valid in the domain of human–computer interaction. In order to answer this question, this hypothesis states that the variables trait reactance, social agency, and involvement are moderators of state reactance, also in the domain of human–computer interaction.*

Dillard and Shen conducted two experiments in which they measured trait reactance and state reactance. Trait reactance was a significant predictor for state reactance in both experiments [7]. Still, it is uncertain, how trait reactance influences state reactance. For example, persons who are high in trait reactance could have a lower threshold for entering a reactance state. Or, persons who are high in trait reactance could have the same threshold as persons with low levels of trait reactance, but their reactance reaction is stronger.

Hypothesis 8.1 *Regarding the relationship of trait reactance and state reactance, one suggestion is, that trait reactance acts as a moderator variable for state reactance. This would mean that persons who show high levels of trait reactance would show stronger effects of reactance restoration, e.g., source derogation. Hypothesis 8.1 therefore, states that trait reactance is a moderator variable for state reactance.*

Apart from trait reactance, there might be further factors that determine if, and to what extent a user will enter a reactant state and how this will affect the consequences for the technical system that caused state reactance. These factors are social agency and involvement.

Hypothesis 8.2 *Roubroeks et al. [19–21] were able to show that a persuasive assistant system could induce state reactance. They also showed that the level of state reactance was higher if the social agency of the assistant was higher. Hypothesis 8.2 therefore states that social agency is a moderator of state reactance.*

Hypothesis 8.3 *Already Brehm stated that the amount of reactance that a person will experience is dependent on how important the threatened freedom is to that person [4, pp. 118]. However, Quick investigated the influence of involvement on the level of state reactance and found no evidence of involvement being of importance [18]. In order to gather more evidence on this topic, Hypothesis 8.3 states, that involvement is a moderator variable of state reactance.*

The laboratory experiment that was conducted to test Hypotheses 7–8.3 is described in the following sections.

11.2 Participants

72 participants participated in the experiment, 40 of which were female. The age of the participants varied between 18 and 40 years with an average age of 27.63 years.

11.3 Methods

A laboratory experiment with a 2×3 conditional partial-between, partial within-participants design was conducted. The between-participants conditions differed in the presence of a robotic head. In one condition, participants interacted with the original Echo device by Amazon.com [1]. In the other condition, participants interacted with a robotic head that was connected to the Echo device. The head moved its lower jaw according to the audio output of Alexa and moved its head and eyes following a control-script (see Sect. 11.3.1). The robot/no-robot condition was introduced to create variance in terms of social agency in the measured data. There were also three within-participants conditions. The between participants condition varied in their persuasion strategy (see Sect. 11.3.2) and was included to create variance in terms of freedom threat.

11.3.1 Robotic Head

A robotic head was constructed by the author of this thesis to serve as an anthropomorphic, social agent to represent Alexa. The robot was based on a full-scale model of a human skull [15]. It included motors to move its eyes in two axes [14], its head in three axes and its lower jaw in one axis. The jaw was connected to a circuit board [5] that allowed for jaw movements that were approximately synchronous to the audio that was played. If the audio signal would increase in amplitude, the jaw would open. If the audio signal would decrease in amplitude, the jaw would close. The frequency of jaw movements could be adjusted by using a moving-average function over the sound signal. The face was modeled via modeling clay. It contained a small, pointy nose and no ears or lips. Also, the eyes were included in a way that they would seem to be rather large. This was done to avoid an uncanny valley effect [16]. Alexa could be connected to the robot head via Bluetooth (Fig. 11.1).

11.3.2 Amazon Alexa Skill

Skills are applications that can be installed for the intelligent personal assistant *Alexa* of Amazon.com [2]. The main task of the participants was to interact with a skill

Fig. 11.1 Setup that was used in the experiment. The robotic head was placed in close proximity to the Echo device. The sound was not emitted directly from the head's position, but from the two speakers placed to the left and to the right of the robot

that represented a to-do list. The skill was implemented by a student in the frame of a master's thesis. The skill was interactive via natural language interaction. It applied two different persuasive strategies to persuade its users into adding more items to the list. One strategy applied the "but you are free to accept or to refuse" technique [9] while arguing for adding a new item. This strategy was used to pose a low threat to the user's freedom. The other strategy was using forceful language without argumentation, posing a high threat to the user's freedom. Additionally, there was a baseline condition where neither of the two strategies was applied. A translated example for the conditions is provided in the following:

> User: *"Add "Congratulate aunt Maria for her birthday" to my list."*
>
> Baseline Skill: *"I have added "Congratulate aunt Maria for her birthday" to your list. Would you like to add" Buy a present for aunt Marias birthday" to your list?"*
>
> Low Threat Skill: *"OK! As requested, I have added "Congratulate aunt Maria for her birthday" to your list. Your aunt would surely be happy about a little present for her birthday! Should I add "buying a present" to your list?"*
>
> High Threat Skill: *"I have added "Congratulate aunt Maria for her birthday" to your list. You have to buy a present for her birthday. Do I have to add "Buying a present" to your list?"*

11.3.3 Procedure

All participants took part in an online survey, several days before they participated in the experiment. The survey was used to measure their level of trait reactance via the questionnaire Fragebogen zur Messung der Psychologischen Reaktanz in the version refined by Herzberg [10, 13].

When the participants took part in the laboratory experiment, they were greeted by the experimenter and led to the laboratory. At first, they received a sheet of instructions and a form of consent. Afterward, the procedure of the experiment was explained shortly. The participants were told that they were to interact with a smart shopping list via a spoken dialogue assistant. The to-do list was able to analyze items and propose new items that correspond to the ones provided by the subjects. They were also asked not to abort any task. Before the actual experiment started, participants went through a trial session to accustom them to the assistant. Afterward, they completed the three within-participants conditions with different persuasion strategies. The assistant would then use either explaining language, imperative language or neutral language when proposing new items. The order of the within-participants conditions was systematically changed to avoid sequence effects. After each condition, the participants completed the RSHCI and several other questionnaires. After all three trials were solved, the participants also filled in a questionnaire for involvement and provided their demographic data.

11.3.3.1 Operationalization

The experiment contained a number of variables that had to be operationalized. There is no consensus for social agency or involvement on how to operationalize these variables. Also, assessment of trait reactance has been a matter of debate.

Trait Reactance

The findings of Hong and Ostini [11] and Tucker et al. [22] pointed toward a psychometric instability of the Fragebogen zur Messung der Psychologischen Reaktanz. On the other hand, Tucket et al. argued, that this instability might be due to the translation into English [22]. Also, Hong and Ostini used this translation [11] when they criticised the questionnaire. In 2002, Herzberg [10] conducted a refinement of the original scale and thereby deleted six items, furthermore, the format was changed from a six-point Likert [12] scale to a four-point Likert scale [10] (see Sect. 2.3.1.3 for further details.). This refined version of the Fragebogen zur Messung der Psychologischen Reaktanz is going to be used to measure trait reactance in the current study.

Social Agency

Social agency was operationalized via the two between-participants conditions of the robotic head (robot/no-robot). This means that there are only two different levels of social agency, which will require to dummy-code the variable for further statistical analysis. The robot can be regarded to increase social agency of the assistant because it is a moving, anthropomorphic head. This is in line with the work by Roubroeks et al. who used a video of a moving cat robot for the high social agency condition [20, 21].

Involvement

Involvement is a concept that is difficult to measure and there is no consensus on how to measure it [3]. In this experiment, involvement was operationalized via the personal habits and preferences of the subjects, regarding the tasks that the Alexa skill was persuading them to. At the end of the experiment, participants received a questionnaire with statements about the topics that were part of the experiment, for example:

> *"Would you like to be reminded to buy presents?"*

If they answered the question with "yes", they got one point for their involvement score, if they answered the question with "no", their involvement score would not rise. The final involvement score was the ratio of "yes" to the total number of statements.

11.4 Results

Hypotheses 7–8.3 are going to be tested using linear regression analysis. The details are presented in the following sections.

11.4.1 Attractiveness Prediction

A series of linear regression analyses is going to be used to investigate Hypothesis 7. Since the proposed model is based on theoretical assumptions and the multiple regression analysis is performed to test these assumptions, the regression method "forced entry" is used. This method includes all independent variables simultaneously [8, p. 322]. For easier interpretation of the regression coefficients, all variables were z-transformed via

$$z = \frac{X - \mu}{\sigma} \tag{11.1}$$

where σ is the standard deviation and μ is the mean of the population.

Attractiveness without State Reactance

The results of the linear regression model using the variables *pragmatic quality* and *hedonic quality* with simultaneous entry as predictors for *attracitiveness* are shown in Table 11.1. The model accounts for a significant amount of variance with corrected $R^2 = 0.635$, F(2,213) = 187.948, p < 0.001.

Table 11.1 Attractiveness prediction without state reactance

	B	SE B	t	p
Constant	5.723E-6	0.041	0	$p = 1$
Pragmatic quality	0.236	0.053	4.493	$p < 0.001$
Hedonic quality	0.631	0.053	12.016	$p < 0.001$

Table 11.2 Attractiveness prediction with state reactance

	B	SE B	t	p
Constant	5.723E-6	0.041	0	$p = 1$
Pragmatic quality	0.132	0.052	2.544	$p < 0.05$
Hedonic quality	0.497	0.054	9.275	$p < 0.001$
State reactance	−0.314	0.052	−5.986	$p < 0.001$

Attractiveness with State Reactance

The results of the hierarchical regression model using the variables pragmatic quality, hedonic quality, and state reactance as predictors for attractiveness are shown in Table 11.2. The model accounts for a significant amount of variance with corrected $R^2 = 0.686$, $F(3,212) = 157.727$, $p < 0.001$.

11.4.2 Moderation Analysis

A moderation analysis following [8, pp. 395] was conducted with the plugin PRO-CESS v2.16.3 for IBM SPSS Statistics 25 to test if the three variables trait reactance, social agency and Involvement influence the effect of state reactance on Attractiveness. Quick [17] reported that trait reactance was correlated with Anger, but not with negative cognitions, suggesting that trait reactance could be a moderator variable for Anger, only. Following this observation and the fact that state reactance is a construct of two components (anger and negative cognitions), the moderator analysis will be performed for these components separately.

The procedure for a moderation analysis using linear regression modeling is as follows: First, all variables need to be standardized and transformed to an interval scale. This includes that the categorical variable social agency had to be dummy-coded. For easier comparison and interpretation, all variables were also z-transformed.

Then, a regression model was calculated that predicts the dependent variable (attractiveness), using the dependent variable (anger or negative cognitions) and the moderator variable as predictors.

In the next step, the interaction effect (e.g., anger × trait reactance) is added as a predictor to the regression model of the previous step.

Finally, the R^2 values are compared. If the R^2 values are significantly different and the interaction effect is a significant predictor, then a moderation effect is shown.

Anger and Trait Reactance

The results of the hierarchical regression model using the variables anger, trait reactance and the interaction effect of anger and trait reactance as predictors for attractiveness are shown in Table 11.3. The model accounts for a significant amount of variance with $R^2 = 0.3996$, F(3,212) = 47.0251, p < 0.000. Inclusion of the interaction effect did not change the amount of explained variance significantly with Δ $R^2 = 0.0036$, F(1,212) = 1.2572, p = 0.2634.

Negative Cognitions and Trait Reactance

The results of the hierarchical regression model using the variables negative cognitions, trait reactance and the interaction effect of negative cognitions and trait reactance as predictors for attractiveness are shown in Table 11.4. The model accounts for a significant amount of variance with $R^2 = 0.5218$, F(3,212) = 77.1006, p < 0.000. Inclusion of the interaction effect did change the amount of explained variance significantly with Δ $R^2 = 0.0184$, F(1,212) = 8.1537, p < 0.05.

Anger and Social Agency

The results of the hierarchical regression model using the variables anger, social agency, and the interaction effect of anger and social agency as predictors for attractiveness are shown in Table 11.5. The model accounts for a significant amount of variance with $R^2 = 0.4758$, F(3,212) = 64.1514, p < 0.001. Inclusion of the interaction effect did change the amount of explained variance significantly with Δ R^2 = 0.0605, F(1,212) = 24.4877, p < 0.001.

Table 11.3 Moderation Effect = Trait Reactance × Anger

	B	SE B	t	p
Constant	−0.0053	−0.0531	−0.0989	p = 0.9213
Trait reactance	0.0627	0.646	1.1633	p = 0.2460
Anger	−0.6283	0.0505	−11.6953	p < 0.001
Moderation effect	0.0480	0.0494	1.1213	p = 0.2634

Table 11.4 Moderation Effect = Trait Reactance × Negative Cognitions

	B	SE B	t	p
Constant	−0.0158	0.0477	−0.3305	p = 0.7383
Trait reactance	0.0794	0.0482	1.6450	p = 0.1723
Negative cognitions	−0.7311	0.0482	−15.1621	p < 0.001
Moderation effect	0.1188	0.0416	2.8555	p < 0.05

Table 11.5 Moderation Effect = Social Agency × Anger

	B	SE B	t	p
Constant	0.0017	0.0496	0.0338	p = 0.9731
Social agency	−0.1548	0.0497	−3.1143	p < 0.01
Anger	−0.6016	0.05	−12.0366	p < 0.001
Moderation effect	0.2478	0.0501	4.9485	p < 0.001

Negative Cognitions and Social Agency

The results of the hierarchical regression model using the variables negative cognitions, social agency, and the interaction effect of negative cognitions and social agency as predictors for attractiveness are shown in Table 11.5. The model accounts for a significant amount of variance with $R^2 = 0.5387$, F(3,212) = 82.5299, p < 0.001. Inclusion of the interaction effect did change the amount of explained variance significantly with $\Delta R^2 = 0.0246$, F(1,212) = 11.3189, p < 0.001 (Table 11.6).

Anger and Involvement

The results of the hierarchical regression model using the variables anger, involvement, and the interaction effect of anger and I involvement as predictors for Attracitiveness are shown in Table 11.7. The model accounts for a significant amount of variance with $R^2 = 0.3965$, F(3,209) = 45.7687, p < 0.001. Inclusion of the interaction effect did not change the amount of explained variance significantly with $\Delta R^2 < 0.0001$, F(1,209) = 0.0089, p = 0.9424.

Table 11.6 Moderation Effect = Social Agency × Negative Cognitions

	B	SE B	t	p
Constant	−0.0024	0.0465	−0.0525	p = 0.9587
Social agency	−0.1407	0.0466	−3.0107	p < 0.001
Negative cognitions	−0.6708	0.0475	−14.1193	p < 0.01
Moderation effect	0.1602	0.0476	3.3644	p < 0.01

Table 11.7 Moderation Effect = Involvement × Anger

	B	SE B	t	p
Constant	−0.0054	00562	−0.0953	p = 0.9241
Involvement	−0.0802	00571	−1.4049	p = 0.1615
Anger	−0.6085	00571	−10.6641	p < 0.001
Moderation effect	0.00053	00565	0.0941	p = 0.9251

Table 11.8 Moderation Effect = Involvement × Negative Cognitions

	B	SE B	t	p
Constant	0.0096	0.052	0.1850	p = 0.8534
Involvement	−0.0569	0.0525	−1.0829	p = 0.2801
Negative cognitions	−0.6849	0.0521	−13.1453	p < 0.001
Moderation effect	−0.0392	0.0579	−0.6778	p = 0.4987

Negative Cognitions and Involvement

The results of the hierarchical regression model using the variables negative cognitions, involvement and the interaction effect of negative cognitions, and involvement as predictors for attractiveness are shown in Table 11.8. The model accounts for a significant amount of variance with $R^2 = 0.4967$, F(3,209) = 68.7552, p < 0.001. Inclusion of the interaction effect did not change the amount of explained variance significantly with $\Delta R^2 = 0.0011$, F(1,209) = 0.4593, p = 0.4987.

Correlations

Pearson correlations were calculated to examine if trait reactance, social agency, or involvement are correlated with either anger or negative cognitions. Neither trait reactance, nor social agency was correlated with anger or negative cognitions. There was a highly significant small positive correlation between involvement and anger with r = 0.279, p < 0.001 and between involvement and negative cognitions with r = 0.289, p < 0.001.

References

1. Amazon.com, Inc.: Amazon echo—black (1st generation) (2018). Retrieved August 19, 2018 from https://www.amazon.com/Amazon-Echo-Bluetooth-Speaker-with-Alexa-Black/dp/B00X4WHP5E
2. Amazon.com: Alexa (2014). Retrieved January 27th, 20117 from https://developer.amazon.com/alexa
3. Andrews, C.J., Durvasula, S.: Suggestions for manipulating and measuring involvement in advertising message content. Adv. Consum. Res. **18**, 194–201 (1991)
4. Brehm, J.W.: A Theory of Psychological Reactance. Academic Press, New York (1966)
5. Computer & Electronic Services—Cowlacious Designs: Scary terry's audio servo driver (2017)
6. Diefenbach, S., Hassenzahl, M.: Handbuch zur Fun-ni Toolbox. Folkwang Universität der Künste, The address of the publisher (2010). Retrieved March 14, 2016 from http://fun-ni.org/wp-content/uploads/Diefenbach+Hassenzahl_2010_HandbuchFun-niToolbox.pdf
7. Dillard, J.P., Shen, L.: On the nature of reactance and its role in persuasive health communication. Commun. Monogr. **72**(2), 144–168 (2005). https://doi.org/10.1080/03637750500111815
8. Field, A.: Discovering Statistics Using IBM SPSS Statistics, vol. 4. SAGE Publications Ltd (2013)
9. Guéguen, N., Pascual, A.: Improving the response rate to a street survey: an evaluation of the "but you are free to accept or to refuse" technique. Psychol. Record **55**(2), 297–303 (2005). https://doi.org/10.1007/BF03395511

10. Herzberg, P.Y.: Zur psychometrischen Optimierung einer Reaktanzskala mittels klassischer IRT-basierter Analysemethoden. Diagnostica **48**(4), 163–171 (2002). https://doi.org/10.1026//0012-1924.48.4.163
11. Hong, S.M., Ostini, R.: Further evaluation of merz's psychological reactance scale. Psychol. Reports **64**(3), 707–710 (1989)
12. Likert, R.: A technique for the measurement of attitudes. Arch, Psychol (1932)
13. Merz, J.: Fragebogen zur messung der psychologischen reaktanz. Diagnostica **29**(1), 75–82 (1983)
14. Monster Guts, LLC.: 2-axis eye kit (2015). Retrieved Juni 21, 2018 from http://www.monsterguts.com/store/product.php?productid=17785
15. Monster Guts, LLC.: 3-axis skull kit (2015). Retrieved Juni 21, 2018 from http://www.monsterguts.com/store/product.php?productid=17782
16. Mori, M.: The uncanny valley. Energy **7**(4), 33–35 (1970)
17. Quick, B.L., Stephenson, M.T.: Examining the role of trait reactance and sensation seeking on perceived threat, state reactance, and reactance restoration. Human Commun. Res. **34**(3), 448–476 (2008). https://doi.org/10.1111/j.1468-2958.2008.00328.x
18. Quick, B.L., Scott, A.M., Ledbetter, A.M.: A close examination of trait reactance and issue involvement as moderators of psychological reactance theory. J. Health Commun. **16**(6), 660–679 (2011). https://doi.org/10.1080/10810730.2011.551989
19. Roubroeks, M., Ham, J., Midden, C.: The dominant robot: threatening robots cause psychological reactance, especially when they have incongruent goals. In: Ploug, T., Hasle, P., Oinas-Kukkonen, H. (eds.) Persuasive Technology, pp. 174–184. Springer, Berlin, Heidelberg (2010)
20. Roubroeks, M., Ham, J., Midden, C.: When artificial social agents try to persuade people: the role of social agency on the occurrence of psychological reactance. Int. J. Soc. Robot. **3**(2), 155–165 (2011). https://doi.org/10.1007/s12369-010-0088-1
21. Roubroeks, M., Midden, C., Ham, J.: Does it make a difference who tells you what to do ? exploring the effect of social agency on psychological reactance. In: Proceedings of the 4th International Conference on Persuasive Technology, Persuasive '09, pp. 15:1–15:6. ACM, New York, NY, USA (2009). https://doi.org/10.1145/1541948.1541970
22. Tucker, R.K., Byers, P.Y.: Factorial validity of merz's psychological reactance scale. Psychol. Reports **61**(3), 811–815 (1987). https://doi.org/10.2466/pr0.1987.61.3.811

Chapter 12
Intermediate Discussion on Determinants of State Reactance

12.1 Smart TV Study

The smart TV study was conducted in order to investigate the influence of system errors on state reactance. To accomplish this, a partial within-partial between-participants design was chosen and three hypothesees were formulated.

12.1.1 Hypothesis 4

The fourth hypothesis claimed that system errors can be a cause for state reactance. To test this hypothesis, two kinds of errors were introduced in the experiment. One, that can be explained by a misinterpretation of the command and one that can not be explained by the users. Both error types were tested against a baseline condition without artificially introduced errors in a within-participants design. Results show significant differences in state reactance between the baseline condition and either of the error conditions. For both conditions with different error types, state reactance was significantly higher, compared to the baseline condition. This is also true for the two components of state reactance, anger (accessed by the factor anger) and negative cognitions (accessed by the factor negative cognitoins). Hypothesis 12.1.1 is therefore confirmed by the experimental results.

12.1.2 Hypothesis 5

The fifth hypothesis claimed that a perceived loss of control over the situation is often the cause of state reactance and that state reactance could be reduced by increasing the user's awareness over the cause of the error. The current experiment used two

© Springer Nature Switzerland AG 2020
P. Ehrenbrink, *The Role of Psychological Reactance in Human–Computer Interaction*, T-Labs Series in Telecommunication Services,
https://doi.org/10.1007/978-3-030-30310-5_12

conditions with artificially introduced errors in a between-participants design to investigate this hypothesis. One condition contained artificially introduced errors where the users could explain the cause of the error as a false interpretation of their voice command by the Smart TV. The other condition contained the same amount of artificially introduced errors but did not provide a hint for the cause of the error. Results show a significantly higher rating in state reactance for the condition where the error cannot be explained, compared to the condition that hints at the reason of the error. Also, the two components of state reactance, Anger, and negative cognitions, were measured to be significantly higher in the condition that did not provide an explanation for the error. Since state reactance was significantly lower in the condition that provided an explanation for the error, Hypothesis 5 can be regarded as confirmed by the experimental results.

12.1.3   Hypothesis 6

Hypothesis 6 claimed that state reactance would result in lower acceptance of the source of the threat to freedom or control. In this experiment, the threat to the user's control was emitted by the smart TV. Acceptance was not directly assessed. Instead, acceptability was measured, because it can be regarded as acceptance potential. In order to evaluate the relationship of state reactance and acceptability, a correlation between the two variables was calculated. Spearman's ρ was 0.695. Like the Pearsons product-moment correlation coefficient, Spearman's ρ can reach values between -1 and 1 [1, pp. 232]. A ρ of 0.695 can, therefore, be regarded as a rather high correlation. Given the apparent correlation, which is also highly significant, Hypothesis 12.1.3 can be regarded as confirmed. However, it needs to be stressed that a correlation does not indicate causality and that the correlation between acceptability and state reactance which could be observed only indicates an association between acceptance and state reactance indirectly. This association also corresponds to Research Question 1, because it has been shown that the relevance of state reactance for human–computer interaction is partly due to its probable nature of having a negative impact on acceptance.

12.1.4   Study Remarks and Weaknesses

Even though the Smart TV was capable of processing natural language commands, the actual commands used by the participants were formulated very function-oriented, e.g., "Show me videos of dogs on YouTube" or "Please show me videos of dogs on YouTube". This was probably also due to priming [5, p. 820], an effect that results in a higher probability of the word being used for voice commands that have been heard or read recently. The tasks for the participants were formulated with the same phrases that could be used in the voice commands, specifically for the purpose of priming the participants so they would use those phrases and thereby produce fewer unexpected errors while operating the Smart TV. Because of the function-oriented formulation, errors that were not artificially introduced were indeed very rare throughout the experiment.

A weakness of the study design is that there was no direct assessment of situation awareness via a self-report measure. Knowledge over the cause of the artificially introduced errors, which would constitute situation awareness, was assumed to be a moderating variable for state reactance and is therefore important for the validity of the results. However, such a measure could have been implemented only at the very end of the experiment because it would have hinted at the artificial nature of the introduced error and therefore compromise the state reactance measurement. It was assumed that the participants would mix up the two conditions if situation awareness would have been assessed at the very end of the experiment, also it was assumed that a self-report assessment of situation awareness is very prone to social desirability and would, therefore, produce unreliable results. Instead, it was chosen to implement a mechanism to strongly hint at the error cause by asking the participants to write down the content of the feedback fields, without separately assessing situation awareness.

Another potential weakness of the study is that freedom threat was not measured. As a confirmatory variable, measurement of freedom threat is not necessary for the Research Question but could have added to the validity of the results. However, the obtained results of the state reactance measurement are quite clear and all hypotheses could be answered without the freedom threat measure.

12.2 Persuasive Assistant Study

The persuasive assistant study was performed to investigate interaction effects between potential moderator variables of state reactance and the two state reactance components anger and negative cognitions. Furthermore, state reactance was used alongside global measures for usability and user experience to estimate the attractiveness of the system.

12.2.1 Hypothesis 7

Hypothesis 7 regarded a potential benefit of using state reactance in human–computer interaction research. It stated that the prediction accuracy of attractiveness, as measured by the AttrakDiff questionnaire can be increased by using state reactance as a predictor, compared to the pragmatic quality and hedonic quality, only. Looking at Table 9.2, it can be seen that the regression coefficient of state reactance is even larger than the one of pragmatic quality, indicating that state reactance is a better predictor than pragmatic quality. In total, the regression model that included state reactance was able to explain $R^2 = 0.691 = 69.1\%$ of the variance of attractiveness, compared to 63.8% without state reactance. Hypothesis 7 can therefore be confirmed.

12.2.2   Hypothesis 8

Hypothesis 8 argued, that moderators of state reactance that are known from domains other than human–computer interaction are also moderators of state reactance for human–computer interaction. To test this hypothesis, several other hypotheses have been formulated. Also, the analysis was performed for the components of state reactance, independently.

Hypothesis 8.1

Hypothesis 8.1 stated that trait reactance would have a moderating effect on state reactance. Moderation analysis came to different results for anger and for negative cognitions. No significant moderation effect of trait reactance could be observed for Anger, however, a significant moderation effect was found for negative cognitions, even though it was not very strong with less than 2% increase of explained variance.

Hypothesis 8.2

Hypothesis 8.2 investigated the moderator variable social agency. It was found to be a significant moderator variable for both components of reactance. The interaction effect with anger accounted for an increase of about 6% of explained variance, for negative cognitions, the interaction effect accounted for about 2% explained variance.

Hypothesis 8.3

No significant interaction effect could be found for involvement.

Hypothesis 8 can partly be confirmed. Two of the three potential moderator variables were observed to have a significant interaction effect with either anger or negative cognitions. It seems, that particularly trait reactance shows an interesting behavior, regarding its role as a moderator variable. Quick and Stephens found a correlation of trait reactance with Anger [4], while the current study found it to have an interaction effect on negative cognitions. These observations could be due to measurement errors. The interaction effect is quite small and moderation only accounts for less than 2% of explained variance. Also, even though the interaction effect was significant, the corresponding confidence level was only at 95%. On the other hand, whether trait reactance has an effect on anger or negative cognitions could depend on the context. In the study by Quick and Stephens, the stimuli were of a health-related context, namely exercising and using sunscreen [4]. The context of the current experiment was rather everyday tasks. Possibly, health-related tasks have a stronger focus on the affective part of the human motivation, while task planning rather addresses human cognition.

12.2.3 Study Remarks and Weaknesses

The persuasive assistant study was conducted to investigate the role of state reactance on the judgment of the users over technical systems. The results confirm findings from the smart TV study, that state reactance is really associated negatively with the judgment of the technical system regarding acceptance and general attractiveness. Also, the influence of potential moderator variables has been investigated. It has been shown that the effect of state reactance on attractiveness can be influenced by social agency, and that this influence acts on both components of state reactance. Also, it was shown that trait reactance only influences the effect of negative cognitions on attractiveness. Trait reactance and social agency can, therefore, be regarded as moderator variables for state reactance. No moderation effect was observed for involvement. However, a correlation analysis showed that there is a small but highly significant positive correlation between involvement and the components of state reactance, suggesting that the level of state reactance might be dependent on involvement, such as suggested by Brehm [2, pp. 118].

After this study had been conducted, Ghazali et al. published a paper describing a similar study [3]. They also found that state reactance levels were rising with higher involvement. However, they observed a direct increase of state reactance with a higher social agency, an effect that was not observed in the current study.

The biggest weakness of the persuasive assistant study is its operationalization of the factor involvement. The involvement score was calculated out of the reported habits of the subjects. In contrast, Ghazali et al. operationalized involvement via a drink that was created either for the participants themselves (high involvement) or for someone else (low involvement) [3]. Such an operationalization is probably more effective in addressing state reactance since the consequences of the two conditions for the participants are more obvious.

References

1. Bortz, J.: Statistik für Human- und Sozialwissenschaftler. Springer (2004)
2. Brehm, J.W.: A Theory of Psychological Reactance. Academic Press, New York (1966)
3. Ghazali, A.S., Ham, J., Barakova, E., Markopoulos, P.: The influence of social cues in persuasive social robots on psychological reactance and compliance. Comput. Hum. Behav. **87**, 58–65 (2018). 10.1016/j.chb.2018.05.016. http://www.sciencedirect.com/science/article/pii/S0747563218302425
4. Quick, B.L., Stephenson, M.T.: Examining the role of trait reactance and sensation seeking on perceived threat, state reactance, and reactance restoration. Hum. Commun. Res. **34**(3), 448–476 (2008). https://doi.org/10.1111/j.1468-2958.2008.00328.x
5. Stadtler, T.: Lexikon der Psychologie. Alfred Kroner Verlag (2013)

Part V
General Discussion

Chapter 13
Research Outcomes

13.1 Research Question 1

The smart home study, described in Chap. 6 was conducted to further investigate if trait reactance is of relevance for human–computer interaction. It was found that different interactions strategies of technical systems can influence how users feel when interacting with these systems, dependent on their level of trait reactance. However, the study failed to show immediate consequences of this for the system itself. The ratings of the system were not influenced. Some evidence that trait reactance can have a direct influence on the system itself was observed in the persuasive assistant study of Chap. 11. It was found that there is an interaction effect of trait reactance with negative cognitions, which is rather small and only significant at a level of $p < 0.05$. Therefore, it is still questionable if trait reactance is of relevance for human–computer interaction.

13.1.1 Research Question 1.1

Interestingly, there is a big discrepancy between the focus of research, which is mostly concentrating on persuasive technology, and the reports by the subjects, who reported situations involving reduced freedom of choice, high effort, and system errors the most.

13.1.2 Research Question 1.2

There is evidence for decreased motivation to use systems that trigger state reactance, in literature, already. In this book, it was shown that state reactance is also correlated with decreased acceptability and perceived attractiveness of a device or service.

© Springer Nature Switzerland AG 2020
P. Ehrenbrink, *The Role of Psychological Reactance in Human–Computer Interaction*, T-Labs Series in Telecommunication Services,
https://doi.org/10.1007/978-3-030-30310-5_13

13.2   Research Question 2

Even though correlation with Dillard and Shen's method of measuring state reactance has been shown, more work on validation of the RSHCI questionnaire is necessary. This includes assessing the scope of domains in which the questionnaire can be used. Steindl et al. [7] stressed that recent research tends to acknowledge state reactance as being a construct composed of not only an affective and a cognitive, but also of a motivational component. RSHCI covers the affective component with its Anger-dimension and the cognitive component with its negative cognitions-dimension. However, the alleged motivational component is only covered partly. RSHCI has the autonomy-dimension that is associated with the user's intention for keeping his or her autonomy.

13.3   Research Question 3

Research Question 3 dealt with the question, which factors influence state reactance in human–computer interaction. This research question was investigated with the two studies described in Chaps. 10 and 11. The smart TV study of Chap. 10 investigated if system errors can cause state reactance. The results showed that system errors can really trigger state reactance and that the effect is higher, if the users do not understand why the error has happened. Other factors that can influence state reactance are the possible moderator variables that have been investigated in Chap. 11. For the moderator variables, Research Question 3.1 was formulated.

13.3.1   Research Question 3.1

The results of the persuasive assistant study of Chap. 11 showed that trait reactance and social agency have are moderator variables of the components of state reactance. While trait reactance only showed an interaction effect with negative cognitions, but not with anger, interaction effects of social agency could be observed with anger and with negative cognitions. No interaction effect of involvement was observed with either of the two-state reactance components.

13.3.2   Research Question 3.2

Research Question 3.2 asked if the factors that influence state reactance in the context of human–computer interaction can also be used to reduce state reactance or the negative consequences of state reactance. Regarding system errors, the results showed that state reactance, as well as negative consequences regarding the rating of the technical system, can be reduced by explaining how the error occurred. Therefore, besides system errors, which are usually undesired by the developers of a technical system, also situation awareness seems to influence state reactance. This implies that even if system errors do appear during the interaction, the level of state reactance that they trigger can be reduced by adequate error handling, which explains the source of the system error and thereby reestablishes the feeling of being in control of the user. Regarding the potential moderator variables that have been investigated in Chap. 11, it can be said that only social agency was observed to have a direct influence on the rating of the technical system (see Tables 11.5 and 11.6): Higher social agency has resulted in a lower attractiveness rating of the system. Neither trait reactance, nor involvement had a significant influence in this respect.

13.4 Methodological Considerations

Operationalization of Trait Reactance

The smart home study that is described in Chap. 6 used a subset of items from Hong's Psychological Reactance Scale to measure trait reactance [3, 4, 8], while the persuasive assistant study from Chap. 11 used Herzberg's refinement of the Fragebogen zur Messung der Psychologischen Reaktanz [2, 6]. Both studies used a refined version of the respective scale, because trait reactance was intended to be analyzed as a single factor. This is in line with other literature that also applied a single-factor measurements [1, 5].

As a psychological construct, trait reactance can probably be split up further into more fine-grained aspects of personality. However, the dimensions of trait reactance that have been identified via the trait reactance questionnaires are quite different and there is no consensus about what such fine-grained aspects of trait reactance would be.

Limitations

One limitation of this thesis is the methodology that was used to identify situations that typically cause state reactance. As already discussed in Sect. 7.5.1, the methodology was not adequate to claim completeness. Also, the students that were participating in the survey were not experts for state reactance. The short introduction to the concept which included two examples that they received, likely biased their reports regarding the situations. However, the survey was mainly conducted to gain a general idea of what situations were prone to trigger state reactance and to derive scenarios for upcoming experiments. The survey was a success, in that respect.

Discriminant Validity of the Reactance Scale for Human–Computer Interaction

It was already discussed in Sect. 9.1, that the RSHCI did not pass the Fornell-Larcker-criterion. One reason for this might have been that the test data lacked variance in terms of stimulus material. The two dimensions of anger and negative cognitions were probably addressed in a similar manner by the presented material, resulting in a high correlation, however, it is possible to include stimuli that only address one of the two dimensions, which should improve variance of the data and reduce the high correlation between the two dimensions. Results from the persuasive assistant study showed, that trait reactance only interacted with negative cognitions in that particular setting, indicating that anger and negative cognitions can indeed be addressed separately.

References

1. Dillard, J.P., Shen, L.: On the nature of reactance and its role in persuasive health communication. Commun. Monogr. **72**(2), 144–168 (2005). https://doi.org/10.1080/03637750500111815
2. Herzberg, P.Y.: Zur psychometrischen Optimierung einer Reaktanzskala mittels klassischer IRT-basierter Analysemethoden. Diagnostica **48**(4), 163–171 (2002). https://doi.org/10.1026//0012-1924.48.4.163
3. Hong, S.M., Faedda, S.: Refinement of the hong psychological reactance scale. Educ. Psychol. Measurement **56**(1), 173–182 (1996). https://doi.org/10.1177/0013164496056001014
4. Hong, S.M., Page, S.: A psychological reactance scale: Development, factor structure and reliability. Psychol. Reports **64**(3_suppl), 1323–1326 (1989). https://doi.org/10.2466/pr0.1989.64.3c.1323
5. Li, C., Meeds, R.: Factors affecting information processing of internet advertisements: a test on exposure conditions, psycholgical reactance, and advertising frequency. In: Proceedings of Conference American Academy of Advertising (Online), pp. 93–101. American Academy of Advertising, Austin (2007)
6. Merz, J.: Fragebogen zur messung der psychologischen reaktanz. Diagnostica **29**(1), 75–82 (1983)
7. Steindl, C., Jonas, E., Sittenthaler, S., Traut-Mattausch, E., Greenberg, J.: Understanding psychological reactance. Zeitschrift für Psychologie **223**, 205–214 (2015). https://doi.org/10.1027/2151-2604/a000222
8. Yost, A.B., Behrend, T.S., Howardson, G., Badger Darrow, J., Jensen, J.M.: Reactance to electronic surveillance: a test of antecedents and outcomes. J. Bus. Psychol. (2018). https://doi.org/10.1007/s10869-018-9532-2

Chapter 14
Conclusion and Future Research

14.1 Conclusion

This thesis asked for the role of psychological reactance in human–computer interaction. The topic has been divided into three Research Question to provide a better overview. The first Research Question asked for whether psychological reactance is of relevance for human–computer interaction, at all. In terms of state reactance, this question could already be answered via a literature search and an expert survey. State reactance is of relevance to human–computer interaction because it can be triggered by technical systems which then has negative consequences, such as decreased ratings and even a refusal to further use the system. Regarding trait reactance, the answer was less clear, only one paper described evidence for trait reactance of being of importance for human–computer interaction. Trait reactance was then investigated in two separate studies. One of these produced small evidence that the trait reactance level of a user can have an influence on the interaction with the system. Ironically, trait reactance seems to be useful as a variable that systems can be adapted to, while the question if this is also reflected in consequences for interaction with the system itself remains inconclusive. Research Question 1.1 then asked in what situations state reactance could occur. The literature search and expert survey resulted in a list of eight situations where state reactance has been investigated or reported by the subjects. One of these, system errors, had not been covered by literature. A subsequent study was performed that closed this gap and showed that system errors can cause state reactance.

Research Question 1.2 regarded the consequences that state reactance of users could have. Literature mostly reported decreases in the motivation to use a service that had triggered state reactance, an observation that was confirmed by the expert survey. Analysis of data from the smart TV study in Chap. 10 and the persuasive assistant study in Chap. 11 then revealed that state reactance is associated with decreased acceptance and can be used as a predictor for the global judgment of a system.

© Springer Nature Switzerland AG 2020

P. Ehrenbrink, *The Role of Psychological Reactance in Human–Computer Interaction*, T-Labs Series in Telecommunication Services,

https://doi.org/10.1007/978-3-030-30310-5_14

An important part of the work described in this thesis is the RSHCI question-naire. Research Question 2 asked how state reactance can be measured adequately in the context of human–computer interaction research. The RSHCI was developed to answer this question. As a questionnaire-only measurement tool, it can be used in laboratory experiments, as well as in online studies without the high danger of social desirability effects, that are associated with measurement techniques that require examiner intervention.

Research Question 3 asked about the factors that influence state reactance. This research question was split up into Research Question 3.1, which specifically asked for moderator variables, and into Research Question 3.2, which asked if factors that influence state reactance can be used to inhibit state reactance effects. The two experiments that were conducted in Part IV were conducted to investigate Research Questions 3–3.2. The results showed that system errors can induce state reactance, and that this effect can be reduced by increasing situation awareness of the users. Regarding the moderator variables of state reactance, evidence suggests that social agency is a moderator variable for both components of state reactance, while trait reactance only influences the effect of negative cognitions. In contrast, no moderation effect of involvement was observed.

In conclusion, it can be said that the role of psychological reactance in human–computer interaction is a two-fold one. State reactance, as a motivational reaction of the user of a technical system to a perceived loss of control, can result in negative consequences for the user's judgment of the system. According to literature, another possible outcome is diminished intention of using the system in the future. Trait reactance, on the other hand, seems to influence the way in which behavior of tech-nical systems is perceived. Highly reactant personalities were shown to experience a freedom threat or a loss of autonomy when interacting with adaptive or proactive devices. Apart from that, trait reactance could not be shown to have a direct effect on a user's judgment of a system.

14.2   Outlook and Future Research

Reactance Scale for Human–Computer Interaction

The RSHCI questionnaire includes a dimension, freedom threat, that is aimed at measuring a person's current need for autonomy. It was intended to be a confirming variable. However, this part of the questionnaire is not jet validated. A validation of the autonomy dimension is one line of future research. On the other hand, it would be beneficial to replace or extend the set of items for autonomy with items that allow for an estimation of the nature of a possible reactance reaction. Reactance will often result in the attempt to restore the lost or threatened freedom by some means. Knowing which strategies will be applied for this restoration attempts can be useful to counteract negative consequences early. Possible strategies for restoring freedom are source derogation and boomerang effects (see Sect. 2.2.1.1).[1] Including items into the RSHCI questionnaire that target these two strategies would therefore greatly increase the benefit of using the questionnaire in practical settings.

Measuring State Reactance without Questionairs

An important step toward the practical use of state reactance as a variable for interaction management is automatic assessment without the use of questionnaires. Automatically assessing state reactance requires some expertise and technical infrastructure, probably including basic text understanding capabilities. Drawing conclusions of a person's level of state reactance could rely on behavioral observations. Similar to early studies on Reactance (refer to Sect. 2.2.1.1), a person's behavior could be monitored and analyzed according to certain criteria. Two behavioral patterns that would have to be looked for are source derogation and boomerang effects. Section 2.2.1.1 provides a detailed description of both behaviors.

Boomerang effects could easily be detected since they are actions that oppose a persuasion attempt. If, for example, an ambient assisted living environment proposes its inhabitant to eat less meat, and soon after, the inhabitant adds sausage and steak to the shopping list, the user model of the environment can assume the inhabitant to be in a state of reactance. Source derogation manifests in a worse opinion about the source of the assumed freedom threat. The fact that, for example, an app makes users reactant could be noticed by bad ratings of the app in its app store. A more privacy-invasive method of detecting source derogation would be analyzing speech of the user while using the app. If users are in social situations and become state reactant because of e.g., an app, they are likely to express their disapprovement to their social partners. Sittenthaler et al. found that the experience of state reactance can be accompanied by an increase in heart rate [1]. It seems however questionable if such a measure could be used to distinguish state reactance from frustration or general arousal. Still, such information could be useful to confirm or support the classification of the user's state.

Reactance Processes

Steindl et al. found that there might be two separate routes for entering a reactant state. An immediate route, which was observed to be applied when participants received illegitimate freedom threats (no reason given for the decline of a renting application), and a delayed route, which followed a legitimate freedom threat (renting application was declined with a valid argument) [2, 3]. Indications for the delayed route were heart rate measures that were conducted immediately after a stimulus and during answering of questionnaires, shortly after the stimulus [2]. In human–computer interaction, such legitimate freedom threats are often planned or considered during system design, e.g., denial of administrative rights for office computers. If there really was a delayed route for state reactance, compliance with such legitimate

---

[1] Source derogation would result in lower acceptance (or ratings) of the source of the freedom threat and a boomerang effect would mean that the person does the opposite of what was intended by the source of the freedom threat.

freedom threats could possibly improve, if countermeasures against state reactance can be applied in time, even after the restriction had to be enforced. There are some such post-action strategies that could be applied in these cases. Legitimate (explained) and illegitimate (not explained) freedom threats have also been investigated in the smart TV study of Chap. 10. The results were interpreted in the way that an explained error increased the situation awareness of the participants and therefore provided a smaller threat to control, resulting in less state reactance. However, a delayed route to state reactance could also explain the observed data in the way that state reactance arousal was not yet at its maximum when it was measured after the legitimate threat.

Inhibition of State Reactance Consequences

state reactance is a motivational state that can result in undesired behaviors of users. These behaviors are attempts to secure or reestablish threatened freedom. Therefore, techniques that reestablish the freedom of users, in a way that is not undesirable from a system perspective, could be used to prevent undesired behavior. Preventing negative consequences of state reactance is a logical next step after it was identified that state reactance can cause problems in human–computer interaction. One approach of inhibiting state reactance has already been investigated in the smart TV study of Chap. 10. It could be shown that increased awareness over the cause of a system error reduced state reactance levels and increased acceptability of the smart TV. Awareness over the cause of an error is helpful for users to solve the problem succeed with their task, thereby it can be regarded as being helpful in reestablishing lost control over the system. Finding similar techniques to reduce effects of state reactance could be of benefit for technical systems in the future. Especially systems that are prone to cause freedom threats, such as persuasive systems or security-related applications, could benefit from applying these techniques.

Context Dependent State Reactance

The study described in Chap. 11 found that trait reactance has a moderation effect on negative cognitions, but not on anger. It was argued in the intermediate discussion in Sect. 12.2.2, that this could be dependent on the context, e.g., that health-related situations rather favor an interaction effect of trait reactance with anger, or that situations that involve more cognitions, such as task planning rather favor an interaction effect of trait reactance with negative cognitions. For future research, it would be interesting to conduct more fine-grained research on how context and different moderator variables influence the expression of anger and negative cognitions. Such information could be useful to develop more effective and less intrusive strategies to avoid state reactance.

References

1. Sittenthaler, S., Jonas, E., Traut-Mattausch, E.: Explaining self and vicarious reactance: a process model approach. Person. Soc. Psychol. Bull. **42**(4), 458–470 (2016). https://doi.org/10.1177/0146167216634055. PMID: 26984012

2. Sittenthaler, S., Traut-Mattausch, E., Steindl, C., Jonas, E.: Salzburger state reactance scale (ssr scale): Validation of a scale measuring state reactance. Zeitschrift für Psychologie **223**, 257–266 (2015). https://doi.org/10.1027/2151-2604/a000227

3. Steindl, C., Jonas, E., Sittenthaler, S., Traut-Mattausch, E., Greenberg, J.: Understanding psychological reactance. Zeitschrift für Psychologie **223**, 205–214 (2015). https://doi.org/10.1027/2151-2604/a000222

Appendix A
Questionnaires

No.	Item (german)	Item (english translation)	Anger	Neg. Cogn.	Threat
1	Wenn das Smart Home mir etwas verbieten will, dann machte ich es erst recht!	If the smart home wants to forbid something, I will do it for sure!			X
2	Vorschläge vom Smart Home sind bei mir willkommen	I welcome suggestions from the smart home			X
3	Das Smart Home darf mir nicht sagen, was ich zu tun habe	The smart home must not tell me what to do			X
4	Ich bin für Ratschläge vom Smart Home immer offen	I am always open to suggestions from the smart home			X
5	Was das Smart Home tut, ist mir völlig egal	I don't care what the smart home is doing			X
6	Ich schätze die Meinung vom Smart Home	I appreciate the smart home's opinion			X
7	**Ich will bestimmen was passiert, nicht das Smart Home!**	**I want to be in control, not the smart home**			**X**
8	Wenn das Smart Home mir etwas sagt, will ich am liebsten das Gegenteil tun	If the smart home tells me something, I would love to do just the opposite			X
9	Ich mache gerne was das Smart Home mir sagt	I like to do what the smart home tells me to do			X
10	Ich fühle mich vom Smart Home eingeschränkt	I feel constricted by the smart home			X
11	Was das Smart Home sagt, ist mir egal	I don't care what the smart home is telling me			X
12	**Ich handle gerne unabhängig vom Smart Home**	**I like to act independently from the smart home**			**X**
13	**Ich möchte nicht, dass das Smart Home mir sagt was ich tun soll**	**I don't want the smart home to tell me what to do**			**X**

(continued)

© Springer Nature Switzerland AG 2020
P. Ehrenbrink, *The Role of Psychological Reactance in Human–Computer Interaction*, T-Labs Series in Telecommunication Services,
https://doi.org/10.1007/978-3-030-30310-5

(continued)

No.	Item (german)	Item (english translation)	Anger	Neg. Cogn.	Threat
14	**Ich lasse mich vom Smart Home nicht bevormunden!**	**I don't let the smart home impose its will on me**			X
15	**Ich allein bestimme was ich tue, nicht das Smart Home**	**I alone determine what I do, not the smart home**			X
16	Der Gedanke vom Smart Home abhängig zu sein, ärgert mich	The thought of being dependent on the smart home irritates me	X		X
17	Ich ärgere mich über das Smart Home	I am annoyed by the smart home	X		
18	**Das Smart Home frustriert mich**	**The smart home frustrates me**	X		
19	**Das Smart Home macht mich wütend**	**The smart home makes me angry**	X		
20	Wenn ich das Smart Home bediene, freue ich mich	I am happy when I operate the smart home	X		
21	Das Smart Home macht mich glücklich	The smart home makes me happy	X		
22	**Ich werde sauer, wenn ich mit dem Smart Home interagieren muss**	**I get mad when I have to interact with the smart home**	X		
23	**Wenn ich das Smart Home nur sehe, platz mir schon der Kragen**	**When I only see that smart home, I burst with anger**	X		
24	Das Smart Home macht mich aggressiv	The smart home makes me aggressive	X		
25	Bei dem Smart Home verliere ich die Geduld	I lose patience with the smart home	X		
26	Es nervt mich, wenn das Smart Home offensichtliche Dinge hervorhebt	I am annoyed if the smart home points out things that are obvious	X		
27	Ich mag das Smart Home nicht	I don't like the smart home		X	
28	Das Smart Home liegt oft falsch	The smart home is often wrong		X	
29	Es freut mich, wenn das Smart Home einen Fehler macht	It makes me happy if the smart home makes a mistake		X	
30	Obwohl das Smart Home seine Aufgabe gut macht, mag ich es nicht	Even though the smart home fulfills its task well, I dislike it		X	
31	**Das Smart Home ist einfach schlecht!**	**The smart home is simply bad!**		X	
32	**Am liebsten würde ich von dem Smart Home nie mehr etwas sehen oder hören**	**At best, I would like to never see or hear anything about that smart home again**		X	
33	Ich mag das Smart Home	I like the smart home		X	
34	**Das Smart Home möchte ich nicht mal geschenkt**	**I don't even want that smart home as a present**		X	
35	**Das ganze Konzept vom Smart Home ist schon im Ansatz mies**	**The whole concept of that smart home is a poor approach**		X	
36	Das Smart Home möchte ich direkt haben	I directly want to have that smart home		X	
37	Wenn ich das Smart Home bediene, fühle ich mich wohl	I feel comfortable when I operate the smart home		X	

Appendix B
Stimuli

Smart TV Study Tasks

1. Stellen Sie den Ton aus. *Turn off the sound.*
   Manipulation: none
2. Lassen Sie sich die Übersicht anzeigen. *Display the overview.*
   Manipulation: none
3. Suchen Sie in der Übersicht nach Actionfilmen *Search for action films.*
   Manipulation: action film → sports films ∨ romantic comedy films
4. Suchen Sie in der Übersicht nach Dokus. *Search the overview for documentaries.*
   Manipulation: documentaries → comedy movie ∨ series
5. Suchen Sie in der Übersicht nach Western aus den 70er Jahren.*Search the overview for western from the 70s.*
   Manipulation: western from the 70s → romantic movie from the 90s ∨ science fiction movie from the 80s
6. Suchen Sie in der Übersicht nach Horrofilmen aus den 80er Jahren.*Search the overview for horror movies from the 80s.*
   Manipulation: horror movies from the 80s → comedy movies from the 90s ∨ cartoons from the 90s
7. Suchen Sie auf YouTube nach Videos zum Thema Kino Trailer. *Search YouTube for videos concerning movie trailers.*
   Manipulation: movie trailers → Let's Play ∨ Minecraft
8. Suchen Sie auf YouTube nach Videos mit Angela Merkel. *Search YouTube for videos concerning Angela Merkel.*
   Manipulation: Angela Markel → Arnold Schwarzenegger ∨ Stefan Raab
9. Lassen Sie sich das Fernsehprogramm für morgen früh anzeigen. *Show the TV schedule for tomorrow morning.*
   Manipulation: tomorrow morning → today evening ∨ the day after tomorrow evening
10. Schränken Sie die Auswahl auf Kindofilme für morgen abend ein. *Only show cinema movies for tomorrow evening.*

© Springer Nature Switzerland AG 2020

P. Ehrenbrink, *The Role of Psychological Reactance in Human–Computer Interaction*, T-Labs Series in Telecommunication Services, https://doi.org/10.1007/978-3-030-30310-5

Manipulation: cinema movies tomorrow evening → series tomorrow evening ∨ news tomorrow evening

11. Lassen Sie sich die Übersicht anzeigen. *Show the overview.*
    Manipulation: overview → YouTube: videos about Australia ∨ football

12. Suchen Sie Filme oder Sendungen mit John Wayne. *Search for movies or series with John Wayne.*
    Manipulation: overview → movies with Sandra Bullock ∨ movies with Harrison Ford

13. Versuchen Sie den Film "Star Wars" zu suchen *Show the movie "Star Wars".*
    Manipulation: Star Wars → Blade Runner ∨ Indiana Jones

14. Lassen Sie sich das Fernsehprogramm vom ZDF anzeigen. *View the program schedule for the channel ZDF.*
    Manipulation: ZDF → ARD ∨ RBB

15. Wechseln Sie in das laufende Fernsehprogramm, d.h. die normale Fernsehansicht. *Switch to the current TV program, the show currently running.*
    Manipulation: none.

16. Schalten Sie auf Pro7.
    Manipulation: none.

17. Stellen Sie den Ton an.
    Manipulation: none.

Appendix C
Other

Dialogue with a Intelligent Personal Assistants

User: *"Hey Siri, what is zero devided by zero?"*

Siri: *"Imagine that you have zero cookies and you split them evenly among zero friends. How many cookies does each person get? See? It doesn't make sense. And Cookie Monster is sad that there are no cookies. And you are sad that you have no friends."*

User: *"Cortana, guess what!"*

Cortana: *"There are 2,335,981,212,665 possible answers to that question."*

© Springer Nature Switzerland AG 2020

P. Ehrenbrink, *The Role of Psychological Reactance in Human–Computer Interaction*, T-Labs Series in Telecommunication Services,

https://doi.org/10.1007/978-3-030-30310-5

Summary

This book investigates the role of psychological reactance in the context of human–computer interaction and its influence on the acceptance of devices and services. Psychological reactance as a construct from social psychology has gained more relevance for the field of human–computer interaction, since the rise of intelligent, ubiquitous systems. In the beginning, this book provides an overview of the state of the art in psychological reactance research, thereby regarding the dualism of the construct as being a personality trait and a motivational state. As the first step of research, a literature search was conducted, which provided an overview of known situations that either triggered state reactance when users interacted with technical systems, or in which trait reactance influenced interaction. Furthermore, a survey was conducted among usability experts, which collected occasions where interaction with technical systems triggered state reactance. The results show that state reactance is of relevance for human–computer interaction, while the relevance of trait reactance could only be shown in a laboratory experiment later on. Further analysis of the data that was collected in the literature search and the expert survey revealed that there is a discrepancy between the common situations in which state reactance was experienced and the situations that have been investigated in literature, up to that point.

Before the role of state reactance for human–computer interaction could be invested further, an adequate measurement tool was needed. The Reactance Scale for Human–Computer Interaction (RSHCI) was designed and validated to measure the two components of state reactance, anger and negative cognitions. The questionnaire was then used in a laboratory experiment to verify if system errors cause state reactance. The occurenece of system errors was the second-most mentioned trigger for state reactance in the expert survey, but no corresponding literature could be found. Results indicate that system errors can cause state reactance, but that this effect can be reduced by increasing awareness of the reason of the system error. The experiment further showed that state reactance is highly correlated with the judgment of the users over system.

© Springer Nature Switzerland AG 2020   141
P. Ehrenbrink, *The Role of Psychological Reactance in Human–Computer Interaction*, T-Labs Series in Telecommunication Services,
https://doi.org/10.1007/978-3-030-30310-5

The final study investigated interaction effects of the components of state reactance with possible moderator variables, regarding the user's global judgment of a technical system. It could be shown that state reactance is a significant predictor for the attractiveness judgment, alongside traditional measures, addressing usability and user experience. It was also shown that trait reactance is a moderator variable for state reactance, in terms of the global judgment of technical systems.

Index

© Springer Nature Switzerland AG 2020
P. Ehrenbrink, *The Role of Psychological Reactance in Human–Computer Interaction*, T-Labs Series in Telecommunication Services,
https://doi.org/10.1007/978-3-030-30310-5

Printed in the United States
By Bookmasters